DRAM CIRCUIT DESIGN

IEEE Press
445 Hoes Lane, P.O. Box 1331
Piscataway, NJ 08855-1331

IEEE Press Editorial Board
Robert J. Herrick, *Editor in Chief*

M. Akay	M. Eden	M. S. Newman
J. B. Anderson	M. E. El-Hawary	M. Padgett
P. M. Anderson	R. F. Hoyt	W. D. Reeve
J. E. Brewer	S. V. Kartalopoulos	G. Zobrist
	D. Kirk	

Kenneth Moore, *Director of IEEE Press*
Catherine Faduska, *Senior Acquisitions Editor*
John Griffin, *Acquisitions Editor*
Robert H. Bedford, *Assistant Acquisitions Editor*
Surendra Bhimani, *Production Editor*

IEEE Solid-State Circuits Society, *Sponsor*
SSC-S Liaison to IEEE Press, Stuart K. Tewksbury

Cover design: William T. Donnelly, *WT Design*
Technical edits and illustrations: Mary J. Miller, *Micron Technology, Inc.*

Technical Reviewers
Joseph P. Skudlarek, *Cypress Semiconductor Corporation, Beaverton, OR*
Roger Norwood, *Micron Technology, Inc., Richardson, TX*
Elizabeth J. Brauer, *Northern Arizona University, Flagstaff, AZ*

Books of Related Interest from the IEEE Press

Electronic and Photonic Circuits and Devices
Edited by Ronald W. Waynant and John K. Lowell
A Selected Reprint Volume
1999   Softcover   232 pp   IEEE Order No. PP5748   ISBN 0-7803-3496-5

EMC and the Printed Circuit Board: Design, Theory, and Layout Made Simple
Mark I. Montrose
1999   Hardcover   344 pp   IEEE Order No. PC5756   ISBN 0-7803-4703-X

High-Performance System Design: Circuits and Logic
Edited by Vojin G. Oklobdzija
A Selected Reprint Volume
1999   Hardcover   560 pp   IEEE Order No. PC 5765   ISBN 0-7803-4716-1

Integrated Circuits for Wireless Communications
Edited by Asad A. Abidi, Paul R. Gray, Robert G. Meyer
A Selected Reprint Volume
1999   Hardcover   688 pp   IEEE Order No. PC5716   ISBN 0-7803-3459-0

CMOS: Circuit Design, Layout, and Simulation
R. Jacob Baker, Harry W. Li, David E. Boyce
1998   Hardcover   944 pp   IEEE Order No. PC5689   ISBN 0-7803-3416-7

DRAM CIRCUIT DESIGN

A Tutorial

Brent Keeth
Micron Technology, Inc.
Boise, Idaho

R. Jacob Baker
Boise State University
Micron Technology, Inc.
Boise, Idaho

IEEE Solid-State Circuits Society, *Sponsor*

Stuart K. Tewksbury and Joe E. Brewer, *Series Editors*

The Institute of Electrical and Electronics Engineers, Inc., New York

This book and other books may be purchased at a discount
from the publisher when ordered in bulk quantities. Contact:

IEEE Press Marketing
Attn: Special Sales
445 Hoes Lane
P.O. Box 1331
Piscataway, NJ 08855-1331
Fax: +1 732 981 9334

For more information about IEEE Press products, visit the
IEEE Online Catalog & Store: http://www.ieee.org/ieeestore.

©2001 by the Institute of Electrical and Electronics Engineers, Inc.
3 Park Avenue, 17th Floor, New York, NY 10016-5997.

*All rights reserved. No part of this book may be reproduced in any form,
nor may it be stored in a retrieval system or transmitted in any form,
without written permission from the publisher.*

Printed in the United States of America.

10   9   8   7   6   5   4   3   2   1

ISBN 0-7803-6014-1
IEEE Order No. PC5863

Library of Congress Cataloging-in-Publication Data

Keeth, Brent, 1960–
   DRAM circuit design : a tutorial / Brent Keeth, R. Jacob Baker.
   p. cm.
   "IEEE Solid-State Circuits Society, sponsor."
   Includes bibliographical references and index.
   ISBN 0-7803-6014-1
   1. Semiconductor storage devices Design and construction. I. Baker, R. Jacob, 1964–
II. Title

TK7895.M4 K425 2000
621.39'732--dc21
                                                              00-059802

For
Susi, John, Katie, Julie, Kyri, Josh,
and the zoo.

Contents

Preface xi

Acknowledgments xiii

List of Figures xv

Chapter 1 An Introduction to DRAM
    1.1 DRAM Types and Operation............................ 1
        1.1.1 The 1k DRAM (First Generation) 1
        1.1.2 The 4k–64 Meg DRAM (Second Generation)........... 7
        1.1.3 Synchronous DRAMs (Third Generation)............. 16
    1.2 DRAM Basics 22
        1.2.1 Access and Sense Operations...................... 26
        1.2.2 Write Operation 30
        1.2.3 Opening a Row (Summary) 31
        1.2.4 Open/Folded DRAM Array Architectures............. 33

Chapter 2 The DRAM Array
    2.1 The Mbit Cell.. 35
    2.2 The Sense Amp 46
        2.2.1 Equilibration and Bias Circuits 46
        2.2.2 Isolation Devices 48
        2.2.3 Input/Output Transistors 49
        2.2.4 Nsense- and Psense-Amplifiers 50
        2.2.5 Rate of Activation 52
        2.2.6 Configurations 52
        2.2.7 Operation 55

    2.3 Row Decoder Elements...............................57
        2.3.1 Bootstrap Wordline Driver 58
        2.3.2 NOR Driver.......................................60
        2.3.3 CMOS Driver 61
        2.3.4 Address Decode Trees............................62
        2.3.5 Static Tree 62
        2.3.6 P&E Tree...63
        2.3.7 Predecoding......................................64
        2.3.8 Pass Transistor Tree 65
    2.4 Discussion..65

Chapter 3 Array Architectures
    3.1 Array Architectures....................................69
        3.1.1 Open Digitline Array Architecture 69
        3.1.2 Folded Array Architecture 79
    3.2 Design Examples: Advanced Bilevel DRAM Architecture.....87
        3.2.1 Array Architecture Objectives......................88
        3.2.2 Bilevel Digitline Construction.......................89
        3.2.3 Bilevel Digitline Array Architecture..................93
        3.2.4 Architectural Comparison..........................98

Chapter 4 The Peripheral Circuitry
    4.1 Column Decoder Elements.............................105
    4.2 Column and Row Redundancy..........................108
        4.2.1 Row Redundancy 110
        4.2.2 Column Redundancy..............................113

Chapter 5 Global Circuitry and Considerations
    5.1 Data Path Elements....................................117
        5.1.1 Data Input Buffer................................117
        5.1.2 Data Write Muxes 121
        5.1.3 Write Driver Circuit 122
        5.1.4 Data Read Path 124
        5.1.5 DC Sense Amplifier (DCSA) 125
        5.1.6 Helper Flip-Flop (HFF)............................127
        5.1.7 Data Read Muxes 129
        5.1.8 Output Buffer Circuit 129
        5.1.9 Test Modes 131

Contents

    5.2 Address Path Elements 132
        5.2.1 Row Address Path 133
        5.2.2 Row Address Buffer............................ 133
        5.2.3 CBR Counter 134
        5.2.4 Predecode Logic............................... 135
        5.2.5 Refresh Rate 135
        5.2.6 Array Buffers 137
        5.2.7 Phase Drivers 137
        5.2.8 Column Address Path........................... 138
        5.2.9 Address Transition Detection..................... 139
    5.3 Synchronization in DRAMs........................... 142
        5.3.1 The Phase Detector............................. 144
        5.3.2 The Basic Delay Element........................ 144
        5.3.3 Control of the Shift Register 145
        5.3.4 Phase Detector Operation........................ 146
        5.3.5 Experimental Results 147
        5.3.6 Discussion 149

Chapter 6 Voltage Converters
    6.1 Internal Voltage Regulators........................... 155
        6.1.1 Voltage Converters............................. 155
        6.1.2 Voltage References............................. 156
        6.1.3 Bandgap Reference............................ 161
        6.1.4 The Power Stage.............................. 162
    6.2 Pumps and Generators 166
        6.2.1 Pumps....................................... 167
        6.2.2 *DVC*2 Generator.............................. 174
    6.3 Discussion 174

Appendix. 177

Glossary 189

Index 193

About the Authors 199

Preface

From the core memory that rocketed into space during the Apollo moon missions to the solid-state memories used in today's commonplace computer, memory technology has played an important, albeit quiet, role during the last century. It has been quiet in the sense that memory, although necessary, is not glamorous and sexy, and is instead being relegated to the role of a commodity. Yet, it is important because memory technology, specifically, CMOS DRAM technology, has been one of the greatest driving forces in the advancement of solid-state technology. It remains a driving force today, despite the segmenting that is beginning to appear in its market space.

The very nature of the commodity memory market, with high product volumes and low pricing, is what ultimately drives the technology. To survive, let alone remain viable over the long term, memory manufacturers must work aggressively to drive down their manufacturing costs while maintaining, if not increasing, their share of the market. One of the best tools to achieve this goal remains the ability for manufacturers to shrink their technology, essentially getting more memory chips per wafer through process scaling. Unfortunately, with all memory manufacturers pursuing the same goals, it is literally a race to see who can get there first. As a result, there is tremendous pressure to advance the state of the art—more so than in other related technologies due to the commodity status of memory.

While the memory industry continues to drive forward, most people can relax and enjoy the benefits—except for those of you who need to join in the fray. For you, the only way out is straight ahead, and it is for you that we have written this book.

The goal of *DRAM Circuit Design: A Tutorial* is to bridge the gap between the introduction to memory design available in most CMOS circuit texts and the advanced articles on DRAM design that are available in technical journals and symposium digests. The book introduces the reader to DRAM theory, history, and circuits in a systematic, tutorial fashion. The level of detail varies, depending on the topic. In most cases, however, our aim is merely to introduce the reader to a functional element and illustrate it with one or more circuits. After gaining familiarity with the purpose and basic operation of a given circuit, the reader should be able to tackle more detailed papers on the subject. We have included a thorough list of papers in the Appendix for readers interested in taking that next step.

The book begins in Chapter 1 with a brief history of DRAM device evolution from the first 1Kbit device to the more recent 64Mbit synchronous devices. This chapter introduces the reader to basic DRAM operation in order to lay a foundation for more detailed discussion later. Chapter 2 investigates the DRAM memory array in detail, including fundamental array circuits needed to access the array. The discussion moves into array architecture issues in Chapter 3, including a design example comparing known architecture types to a novel, stacked digitline architecture. This design example should prove useful, for it delves into important architectural trade-offs and exposes underlying issues in memory design. Chapter 4 then explores peripheral circuits that support the memory array, including column decoders and redundancy. The reader should find Chapter 5 very interesting due to the breadth of circuit types discussed. This includes data path elements, address path elements, and synchronization circuits. Chapter 6 follows with a discussion of voltage converters commonly found on DRAM designs. The list of converters includes voltage regulators, voltage references, $V_{DD}/2$ generators, and voltage pumps. We wrap up the book with the Appendix, which directs the reader to a detailed list of papers from major conferences and journals.

Brent Keeth
R. Jacob Baker

Acknowledgments

We acknowledge with thanks the pioneering work accomplished over the past 30 years by various engineers, manufacturers, and institutions that have laid the foundation for this book. Memory design is no different than any other field of endeavor in which new knowledge is built on prior knowledge. We therefore extend our gratitude to past, present, and future contributors to this field. We also thank Micron Technology, Inc., and the high level of support that we received for this work. Specifically, we thank the many individuals at Micron who contributed in various ways to its completion, including Mary Miller, who gave significant time and energy to build and edit the manuscript, and Jan Bissey and crew, who provided the wonderful assortment of SEM photographs used throughout the text.

Brent Keeth
R. Jacob Baker

List of Figures

Chapter 1 An Introduction to DRAM

1.1	1,024-bit DRAM functional diagram	2
1.2	1,024-bit DRAM pin connections	3
1.3	Ideal address input buffer	3
1.4	Layout of a 1,024-bit memory array	4
1.5	1k DRAM Read cycle	5
1.6	1k DRAM Write cycle	6
1.7	1k DRAM Refresh cycle	6
1.8	3-transistor DRAM cell	7
1.9	Block diagram of a 4k DRAM	9
1.10	4,096-bit DRAM pin connections	9
1.11	Address timing	10
1.12	1-transistor, 1-capacitor (1T1C) memory cell	11
1.13	Row of N dynamic memory elements	12
1.14	Page mode	15
1.15	Fast page mode	16
1.16	Nibble mode	16
1.17	Pin connections of a 64Mb SDRAM with 16-bit I/O	17
1.18	Block diagram of a 64Mb SDRAM with 16-bit I/O	19
1.19	SDRAM with a latency of three	21

xv

1.20	Mode register	23
1.21	1T1C DRAM memory cell	24
1.22	Open digitline memory array schematic	25
1.23	Open digitline memory array layout	25
1.24	Simple array schematic	26
1.25	Cell access waveforms	27
1.26	DRAM charge-sharing	28
1.27	Sense amplifier schematic	28
1.28	Sensing operation waveforms	29
1.29	Sense amplifier schematic with I/O devices	30
1.30	Write operation waveforms	31
1.31	A folded DRAM array	33

Chapter 2 The DRAM Array

2.1	Mbit pair layout	36
2.2	Layout to show array pitch	36
2.3	Layout to show $8F^2$ derivation	38
2.4	Folded digitline array schematic	39
2.5	Digitline twist schemes	40
2.6	Open digitline array schematic	41
2.7	Open digitline array layout	42
2.8	Buried capacitor cell process cross section	43
2.9	Buried capacitor cell process SEM image	43
2.10	Buried digitline mbit cell layout	43
2.11	Buried digitline mbit process cross section	44
2.12	Buried digitline mbit process SEM image	45
2.13	Trench capacitor mbit process cross section	45
2.14	Trench capacitor mbit process SEM image	46
2.15	Equilibration schematic	47

List of Figures xvii

2.16	Equilibration and bias circuit layout	48
2.17	I/O transistors	50
2.18	Basic sense amplifier block	51
2.19	Standard sense amplifier block	53
2.20	Complex sense amplifier block	53
2.21	Reduced sense amplifier block	53
2.22	Minimum sense amplifier block	54
2.23	Single-metal sense amplifier block	55
2.24	Waveforms for the Read-Modify-Write cycle	56
2.25	Bootstrap wordline driver	59
2.26	Bootstrap operation waveforms	59
2.27	Donut gate structure layout	60
2.28	NOR driver	61
2.29	CMOS driver	62
2.30	Static decode tree	63
2.31	P&E decode tree	64
2.32	Pass transistor decode tree	65

Chapter 3 Array Architectures

3.1	Open digitline architecture schematic	71
3.2	Open digitline 32-Mbit array block	73
3.3	Single-pitch open digitline architecture	76
3.4	Open digitline architecture with dummy arrays	77
3.5	Folded digitline array architecture schematic	80
3.6	Folded digitline architecture 32-Mbit array block	84
3.7	Development of bilevel digitline architecture	90
3.8	Digitline vertical twisting concept	90
3.9	Bilevel digitline architecture schematic	92
3.10	Vertical twisting schemes	93

3.11	Plaid 6F^2 mbit array	94
3.12	Bilevel digitline array schematic	95
3.13	Bilevel digitline architecture 32-Mbit array block	97

Chapter 4 The Peripheral Circuitry

4.1	Column decode	107
4.2	Column selection timing	107
4.3	Column decode: P&E logic	109
4.4	Column decode waveforms	110
4.5	Row fuse block	112
4.6	Column fuse block	115
4.7	8-Meg x 8-sync DRAM poly fuses	116

Chapter 5 Global Circuitry and Considerations

5.1	Data input buffer	118
5.2	Stub series terminated logic (SSTL)	119
5.3	Differential amplifier-based input receiver	120
5.4	Self-biased differential amplifier-based input buffer	121
5.5	Fully differential amplifier-based input buffer	121
5.6	Data Write mux	123
5.7	Write driver	124
5.8	*I/O* bias circuit	125
5.9	*I/O* bias and operation waveforms	126
5.10	DC sense amp	127
5.11	DCSA operation waveforms	128
5.12	A helper flip-flop	128
5.13	Data Read mux	130
5.14	Output buffer	131
5.15	Row address buffer	134
5.16	Row address predecode circuits	136

List of Figures   xix

5.17	Phase decoder/driver	138
5.18	Column address buffer	139
5.19	Equilibration driver	140
5.20	Column predecode logic	141
5.21	SDRAM *CLK* input and *DQ* output	142
5.22	Block diagram for DDR SDRAM DLL	143
5.23	Data timing chart for DDR DRAM	144
5.24	Phase detector used in RSDLL	145
5.25	Symmetrical delay element used in RSDLL	146
5.26	Delay line and shift register for RSDLL	146
5.27	Measured rms jitter versus input frequency	148
5.28	Measured delay per stage versus V_{CC} and temperature	148
5.29	Measured *ICC* versus input frequency	149
5.30	Two-way arbiter as a phase detector	150
5.31	Circuit for generating shift register control	150
5.32	A double inverter used as a delay element	151
5.33	Transmission gates added to delay line	151
5.34	Inverter implementation	152
5.35	Segmenting delays for additional clocking taps	152

Chapter 6 Voltage Converters

6.1	Ideal regulator characteristics	157
6.2	Alternative regulator characteristics	158
6.3	Resistor/diode voltage reference	159
6.4	Voltage regulator characteristics	159
6.5	Improved voltage reference	161
6.6	Bandgap reference circuit	162
6.7	Power op-amp	163
6.8	Power stage	164

6.9	Regulator control logic	166
6.10	Simple voltage pump circuit	168
6.11	V_{CCP} pump	169
6.12	V_{BB} pump	169
6.13	Ring oscillator	169
6.14	V_{CCP} regulator	170
6.15	V_{BB} regulator	170
6.16	V_{CCP} differential regulator	173
6.17	V_{BB} differential regulator	174
6.18	Simple $DVC2$ generator	175
6.19	$DVC2$ generator	176

Chapter 1

An Introduction to DRAM

Dynamic random access memory (DRAM) integrated circuits (ICs) have existed for more than twenty-five years. DRAMs evolved from the earliest 1-kilobit (Kb) generation to the recent 1-gigabit (Gb) generation through advances in both semiconductor process and circuit design technology. Tremendous advances in process technology have dramatically reduced feature size, permitting ever higher levels of integration. These increases in integration have been accompanied by major improvements in component yield to ensure that overall process solutions remain cost-effective and competitive. Technology improvements, however, are not limited to semiconductor processing. Many of the advances in process technology have been accompanied or enabled by advances in circuit design technology. In most cases, advances in one have enabled advances in the other. In this chapter, we introduce some fundamentals of the DRAM IC, assuming that the reader has a basic background in complementary metal-oxide semiconductor (CMOS) circuit design, layout, and simulation [1].

1.1 DRAM TYPES AND OPERATION

To gain insight into how modern DRAM chips are designed, it is useful to look into the evolution of DRAM. In this section, we offer an overview of DRAM types and modes of operation.

1.1.1 The 1k DRAM (First Generation)

We begin our discussion by looking at the 1,024-bit DRAM (1,024 x 1 bit). Functional diagrams and pin connections appear in Figure 1.1 and Figure 1.2, respectively. Note that there are 10 address inputs with pin labels R_1–R_5 and C_1–C_5. Each address input is connected to an on-chip address input buffer. The input buffers that drive the *row (R)* and *column*

1

(C) decoders in the block diagram have two purposes: to provide a known *input capacitance* (C_{IN}) on the address input pins and to detect the input address signal at a known level so as to reduce timing errors. The level V_{TRIP}, an idealized trip point around which the input buffers slice the input signals, is important due to the finite transition times on the chip inputs (Figure 1.3). Ideally, to avoid distorting the duration of the logic zeros and ones, V_{TRIP} should be positioned at a known level relative to the maximum and minimum input signal amplitudes. In other words, the reference level should change with changes in temperature, process conditions, *input maximum amplitude (V_{IH})*, and *input minimum amplitude (V_{IL})*. Having said this, we note that the input buffers used in first-generation DRAMs were simply inverters.

Continuing our discussion of the block diagram shown in Figure 1.1, we see that five address inputs are connected through a decoder to the 1,024-bit memory array in both the row and column directions. The total number of addresses in each direction, resulting from decoding the 5-bit word, is 32. The single memory array is made up of 1,024 memory elements laid out in a square of 32 rows and 32 columns. Figure 1.4 illustrates the conceptual layout of this memory array. A memory element is located at the intersection of a row and a column.

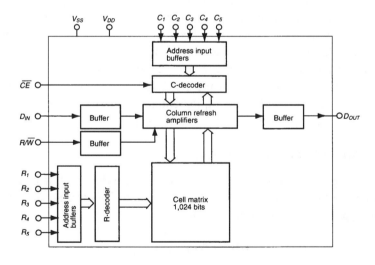

Figure 1.1 1,024-bit DRAM functional diagram.

Sec. 1.1 DRAM Types and Operation

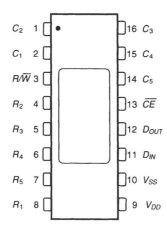

Figure 1.2 1,024-bit DRAM pin connections.

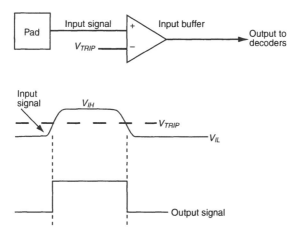

Figure 1.3 Ideal address input buffer.

By applying an address of all zeros to the 10 address input pins, the memory data located at the intersection of row 0, $RA\,0$, and column 0, $CA\,0$, is accessed. (It is either written to or read out, depending on the state of the $R/\overline{W}$ input and assuming that the $\overline{CE}$ pin is LOW so that the chip is enabled.)

It is important to realize that a single bit of memory is accessed by using both a row and a column address. Modern DRAM chips reduce the number of external pins required for the memory address by using the same pins for

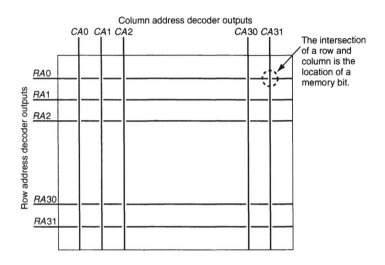

Figure 1.4 Layout of a 1,024-bit memory array.

both the row and column address inputs *(address multiplexing)*. A clock signal *row address strobe* $(\overline{RAS})$ strobes in a row address and then, on the same set of address pins, a clock signal *column address strobe* $(\overline{CAS})$ strobes in a column address at a different time.

Also note how a first-generation memory array is organized as a logical square of memory elements. (At this point, we don't know what or how the memory elements are made. We just know that there is a circuit at the intersection of a row and column that stores a single bit of data.) In a modern DRAM chip, many smaller memory arrays are organized to achieve a larger memory size. For example, 1,024 smaller memory arrays, each composed of 256 kbits, may constitute a 256-Meg (256 million bits) DRAM.

1.1.1.1 Reading Data Out of the 1k DRAM. Data can be read out of the DRAM by first putting the chip in the Read mode by pulling the $R/\overline{W}$ pin HIGH and then placing the chip enable pin $\overline{CE}$ in the LOW state. Figure 1.5 illustrates the timing relationships between changes in the address inputs and data appearing on the D_{OUT} pin. Important timing specifications present in this figure are *Read cycle time* (t_{RC}) and *Access time* (t_{AC}). The term t_{RC} specifies how fast the memory can be read. If t_{RC} is 500 ns, then the DRAM can supply 1-bit words at a rate of 2 MHz. The term t_{AC} speci-

Sec. 1.1 DRAM Types and Operation

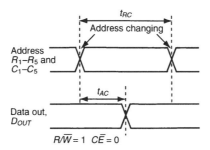

Figure 1.5 1k DRAM Read cycle.

fies the maximum length of time after the input address is changed before the output data (D_{OUT}) is valid.

1.1.1.2 Writing to the 1k DRAM. Writing data to the DRAM is accomplished by bringing the $R/\overline{W}$ input LOW with valid data present on the D_{IN} pin. Figure 1.6 shows the timing diagram for a Write cycle. The term *Write cycle time (t_{WC})* is related to the maximum frequency at which we can write data into the DRAM. The term *Address to Write delay time (t_{AW})* specifies the time between the address changing and the $R/\overline{W}$ input going LOW. Finally, *Write pulse width (t_{WP})* specifies how long the input data must be present before the $R/\overline{W}$ input can go back HIGH in preparation for another Read or Write to the DRAM. When writing to the DRAM, we can think of the $R/\overline{W}$ input as a clock signal.

1.1.1.3 Refreshing the 1k DRAM. The dynamic nature of DRAM requires that the memory be refreshed periodically so as not to lose the contents of the memory cells. Later we will discuss the mechanisms that lead to the dynamic operation of the memory cell. At this point, we discuss how memory Refresh is accomplished for the 1k DRAM.

Refreshing a DRAM is accomplished internally: external data to the DRAM need not be applied. To refresh the DRAM, we periodically access the memory with every possible row address combination. A timing diagram for a Refresh cycle is shown in Figure 1.7. With the $\overline{CE}$ input pulled HIGH, the address is changed, while the $R/\overline{W}$ input is used as a strobe or clock signal. Internally, the data is read out and then written back into the same location at full voltage; thus, logic levels are restored (or refreshed).

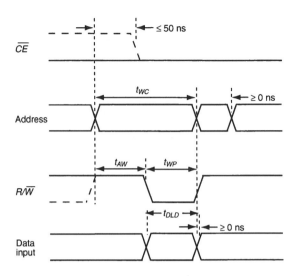

Figure 1.6 1k DRAM Write cycle.

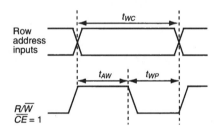

Figure 1.7 1k DRAM Refresh cycle.

1.1.1.4 A Note on the Power Supplies. The voltage levels used in the 1k DRAM are unusual by modern-day standards. In reviewing Figure 1.2, we see that the 1k DRAM chip uses two power supplies: V_{DD} and V_{SS}. To begin, V_{SS} is a greater voltage than V_{DD}: V_{SS} is nominally 5 V, while V_{DD} is –12 V. The value of V_{SS} was set by the need to interface to logic circuits that were implemented using transistor-transistor logic (TTL) logic. The 17-V difference between V_{DD} and V_{SS} was necessary to maintain a large signal-to-noise ratio in the DRAM array. We discuss these topics in greater detail later in the book. The V_{SS} power supply used in modern DRAM designs, at the time of this writing, is generally zero; the V_{DD} is in the neighborhood of 2.5 V.

Sec. 1.1 DRAM Types and Operation

1.1.1.5 The 3-Transistor DRAM Cell. One of the interesting circuits used in the 1k DRAM (and a few of the 4k and 16k DRAMs) is the 3-transistor DRAM memory cell shown in Figure 1.8. The column- and rowlines shown in the block diagram of Figure 1.1 are split into Write and Read line pairs. When the Write rowline is HIGH, M1 turns ON. At this point, the data present on the Write columnline is passed to the gate of M2, and the information voltage charges or discharges the input capacitance of M2. The next, and final, step in writing to the mbit cell is to turn OFF the Write rowline by driving it LOW. At this point, we should be able to see why the memory is called dynamic. The charge stored on the input capacitance of M2 will leak off over time.

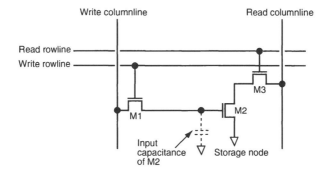

Figure 1.8 3-transistor DRAM cell.

If we want to read out the contents of the cell, we begin by first precharging the Read columnline to a known voltage and then driving the Read rowline HIGH. Driving the Read rowline HIGH turns M3 ON and allows M2 either to pull the Read columnline LOW or to not change the precharged voltage of the Read columnline. (If M2's gate is a logic LOW, then M2 will be OFF, having no effect on the state of the Read columnline.) The main drawback of using the 3-transistor DRAM cell, and the reason it is no longer used, is that it requires two pairs of column and rowlines and a large layout area. Modern 1-transistor, 1-capacitor DRAM cells use a single rowline, a single columnline, and considerably less area.

1.1.2 The 4k–64 Meg DRAM (Second Generation)

We distinguish second-generation DRAMs from first-generation DRAMs by the introduction of multiplexed address inputs, multiple memory arrays, and the 1-transistor/1-capacitor memory cell. Furthermore, second-generation DRAMs offer more modes of operation for greater flexibility or higher speed operation. Examples are *page mode, nibble mode, static column mode, fast page mode* (FPM), and *extended data out* (EDO). Second-generation DRAMs range in size from 4k (4,096 x 1 bit, i.e., 4,096 address locations with 1-bit input/output word size) up to 64 Meg (67,108,864 bits) in memory sizes of 16 Meg x 4 organized as 16,777,216 address locations with 4-bit input/output word size, 8 Meg x 8, or 4 Meg x 16.

Two other major changes occurred in second-generation DRAMs: (1) the power supply transitioned to a single 5 V and (2) the technology advanced from NMOS to CMOS. The change to a single 5 V supply occurred at the 64kbit density. It simplified system design to a single power supply for the memory, processor, and any TTL logic used in the system. As a result, rowlines had to be driven to a voltage greater than 5 V to turn the NMOS access devices fully ON (more on this later), and the substrate held at a potential less than zero. For voltages outside the supply range, charge pumps are used (see Chapter 6). The move from NMOS to CMOS, at the 1Mb density level, occurred because of concerns over speed, power, and layout size. At the cost of process complexity, complementary devices improved the design.

1.1.2.1 Multiplexed Addressing. Figure 1.9 shows a 4k DRAM block diagram, while Figure 1.10 shows the pin connections for a 4k chip. Note that compared to the block diagram of the 1k DRAM shown in Figure 1.1, the number of address input pins has decreased from 10 to 6, even though the memory size has quadrupled. This is the result of using multiplexed addressing in which the same address input pins are used for both the row and column addresses. The row address strobe $(\overline{RAS})$ input clocks the address present on the DRAM address pins A_0 to A_5 into the row address latches on the falling edge. The column address strobe $(\overline{CAS})$ input clocks the input address into the column address latches on its falling edge.

Figure 1.11 shows the timing relationships between $\overline{RAS}$, $\overline{CAS}$, and the address inputs. Note that t_{RC} is still (as indicated in the last section) the random cycle time for the DRAM, indicating the maximum rate we can write to or read from a DRAM. Note too how the row (or column) address must be present on the address inputs when $\overline{RAS}$ (or $\overline{CAS}$) goes LOW. The parameters t_{RAS} and t_{CAS} indicate how long $\overline{RAS}$ or $\overline{CAS}$ must remain LOW after clocking in a column or row address. The parameters t_{ASR}, t_{RAH}, t_{ASC}, and t_{CAH} indicate the setup and hold times for the row and column addresses, respectively.

Sec. 1.1 DRAM Types and Operation

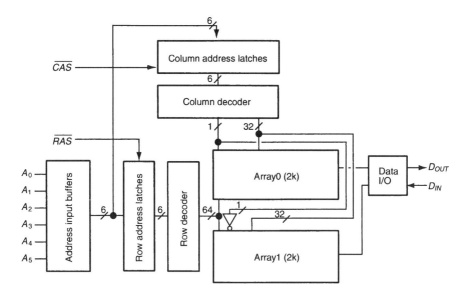

Figure 1.9 Block diagram of a 4k DRAM.

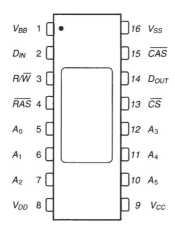

Figure 1.10 4,096-bit DRAM pin connections.

10   Chap. 1   An Introduction to DRAM

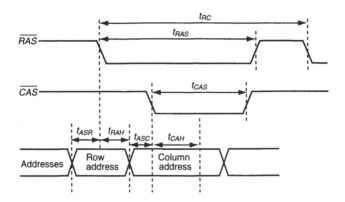

Figure 1.11 Address timing.

1.1.2.2 Multiple Memory Arrays. As mentioned earlier, second-generation DRAMs began to use multiple or segmented memory arrays. The main reason for splitting up the memory into more than one array at the cost of a larger layout area can be understood by considering the parasitics present in the dynamic memory circuit element. To understand the origins of these parasitics, consider the modern DRAM memory cell comprising one MOSFET and one capacitor, as shown in Figure 1.12.

In the next section, we cover the operation of this cell in detail. Here we introduce the operation of the cell. Data is written to the cell by driving the rowline (a.k.a., wordline) HIGH, turning ON the MOSFET, and allowing the columnline (a.k.a., digitline or bitline) to charge or discharge the storage capacitor. After looking at this circuit for a moment, we can make the following observations:

1. The wordline (rowline) may be fabricated using polysilicon (poly). This allows the MOSFET to be formed by crossing the poly wordline over an n+ active area.
2. To write a full V_{CC} logic voltage (where V_{CC} is the maximum positive power supply voltage) to the storage capacitor, the rowline must be driven to a voltage greater than V_{CC} + the n-channel MOSFET threshold voltage (with body effect). This voltage, $> V_{CC} + V_{TH}$, is often labeled V_{CC} *pumped* (V_{CCP}).
3. The bitline (columnline) may be made using metal or polysilicon. The main concern, as we'll show in a moment, is to reduce the parasitic capacitance associated with the bitline.

Sec. 1.1 DRAM Types and Operation   11

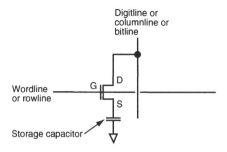

Figure 1.12 1-transistor, 1-capacitor (1T1C) memory cell.

Consider the row of N dynamic memory elements shown in Figure 1.13. Typically, in a modern DRAM, N is 512, which is also the number of bitlines. When a row address is strobed into the DRAM, via the address input pins using the falling edge of $\overline{RAS}$, the address is decoded to drive a wordline (rowline) to V_{CCP}. *This turns ON an entire row in a DRAM memory array.* Turning ON an entire row in a DRAM memory array allows the information stored on the capacitors to be sensed (for a Read) via the bitlines or allows the charging or discharging, via the bitlines, of the storage capacitors (for a Write). Opening a row of data by driving a wordline HIGH is a *very important* concept for understanding the modes of DRAM operation. For Refresh, we only need to supply row addresses during a Refresh operation. For page Reads—when a row is open—a large amount of data, which is set by the number of columns in the DRAM array, can be accessed by simply changing the column address.

We're now in a position to answer the question: "Why are we limited to increasing the number of columnlines (or bitlines) used in a memory array?" or "Why do we need to break up the memory into smaller memory arrays?" The answer to these questions comes from the realization that the more bitlines we use in an array, the longer the delay through the wordline (see Figure 1.13).

If we drive the wordline on the left side of Figure 1.13 HIGH, the signal will take a finite time to reach the end of the wordline (the wordline on the right side of Figure 1.13). This is due to the distributed resistance/capacitance structure formed by the resistance of the polysilicon wordline and the capacitance of the MOSFET gates. The delay limits the speed of DRAM operation. To be precise, it limits how quickly a row can be opened and closed. To reduce this *RC* time, a polycide wordline is formed by adding a silicide, for example, a mixture of a refractory metal such as tungsten with

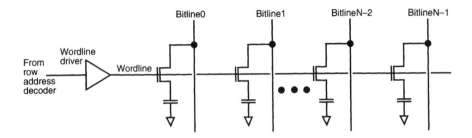

Figure 1.13 Row of N dynamic memory elements.

polysilicon, on top of polysilicon. Using a polycide wordline will have the effect of reducing the wordline resistance. Also, additional drivers can be placed at different locations along the wordline, or the wordline can be *stitched* at various locations with metal.

The limitations on the additional number of wordlines can be understood by realizing that by adding more wordlines to the array, more parasitic capacitance is added to the bitlines. This parasitic capacitance becomes important when sensing the value of data charge stored in the memory element. We'll discuss this in more detail in the next section.

1.1.2.3 Memory Array Size. A comment is in order about memory array size and how addressing can be used for setting word and page size. (We'll explain what this means in a moment.) If we review the block diagram of the 4k DRAM shown in Figure 1.9, we see that two 2k-DRAM memory arrays are used. Each 2k memory is composed of 64 wordlines and 32 bitlines for 2,048 memory elements/address locations per array. In the block diagram, notice that a single bit, coming from the column decoder, can be used to select data, via the bitlines, from Array0 or Array1.

From our discussion earlier, we can open a row in Array0 while at the same time opening a row in Array1 by simply applying a row address to the input address pins and driving $\overline{RAS}$ LOW. Once the rows are open, it is a simple matter of changing the column address to select different data associated with the same open row from either array. If our word size is 1 bit, we could define a page as being 64 bits in length (32 bits from each array). We could also define our page size as 32 bits with a 2-bit word for input/output. We would then say that the DRAM is a 4k DRAM organized as 2k x 2. Of course, in the 4k DRAM, in which the number of bits is small, the concepts of page reads or size aren't too useful. We present them here simply to illustrate the concepts. Let's consider a more practical and modern configuration.

Sec. 1.1 DRAM Types and Operation 13

Suppose we have a 64-Meg DRAM organized as 16 Meg x 4 (4 bits input/output) using 4k row address locations and 4k column address locations (12 bits or pins are needed for each 4k of addressing). If our (sub) memory array size is 256kbits, then we have a total of 256 memory arrays on our DRAM chip. We'll assume that there are 512 wordlines and 512 bitlines (digitlines), so that the memory array is logically square. (However, physically, as we shall see, the array is not square.) Internal to the chip, in the address decoders, we can divide the row and column addresses into two parts: the lower 9 bits for addressing the wordlines/bitlines in a 256k memory array and the upper 3 bits for addressing one of the 64 "group-of-four" memory arrays (6 bits total coming from the upper 3 bits of the row and column addresses).

Our 4-bit word comes from the group-of-four memory arrays (one bit from each memory array). We can define a page of data in the DRAM by realizing that when we open a row in each of the four memory arrays, we are accessing 2k of data (512 bits/array x 4 arrays). By simply changing the column address without changing the row address and thus opening another group-of-four wordlines, we can access the 2k "page" of data. With a little imagination, we can see different possibilities for the addressing. For example, we could open 8 group-of-four memory arrays with a row address and thus increase the page size to 16k, or we could use more than one bit at a time from an array to increase word size.

1.1.2.4 Refreshing the DRAM. Refreshing the DRAM is accomplished by sequentially opening each row in the DRAM. (We'll discuss how the DRAM cell is refreshed in greater detail later in the book.) If we use the 64-Meg example in the last section, we need to supply 4k row addresses to the DRAM by changing the external address inputs from 000000000000 to 111111111111 while clocking the addresses into the DRAM using the falling edge of $\overline{RAS}$. In some DRAMs, an internal row address counter is present to make the DRAM easier to refresh. The general specification for 64-Meg DRAM Refresh is that all rows must be refreshed at least every 64 ms, which is an average of 15.7 μs per row. This means, that if the Read cycle time t_{RC} is 100 ns (see Figure 1.11), it will take 4,096 • 100 ns or 410 μs to refresh a DRAM with 4k of row addresses. The percentage of time the DRAM is unavailable due to Refresh can be calculated as 410 μs/64 ms or 0.64% of the time. Note that the Refresh can be a burst, taking 410 μs as just described, or distributed, where a row is refreshed every 15.7 μs.

1.1.2.5 Modes of Operation. From the last section, we know that we can open a row in one or more DRAM arrays concurrently, allowing a page of data to be written to or read from the DRAM. In this section, we look at the different modes of operation possible for accessing this data via the column address decoder. Our goal in this section is not to present all possible modes of DRAM operation but rather to discuss the modes that have been used in second-generation DRAMs. These modes are page mode, nibble mode, static column mode, fast page mode, and extended data out.

Figure 1.14 shows the timing diagram for a page mode Read, Write, and Read-Modify-Write. We can understand this timing diagram by first noticing that when $\overline{RAS}$ goes LOW, we clock in a row address, decode the row address, and then drive a wordline in one or more memory arrays to V_{CCP}. The result is an open row(s) of data sitting on the digitlines (columnlines). *Only one row can be opened in any single array at a time.* Prior to opening a row, the bitlines are precharged to a known voltage. (Precharging to $V_{CC}/2$ is typically performed using internal circuitry.) Also notice at this time that data out, D_{OUT}, is in a Hi-Z state; that is, the DRAM is not driving the bus line connected to the D_{OUT} pin.

The next significant timing event occurs when $\overline{CAS}$ goes LOW and the column address is clocked into the DRAM (Figure 1.14). At this time, the column address is decoded, and, assuming that the data from the open row is sitting on the digitlines, it is steered using the column address decoder to D_{OUT}. We may have an open row of 512 bits, but we are steering only one bit to D_{OUT}. Notice that when $\overline{CAS}$ goes HIGH, D_{OUT} goes back to the Hi-Z state.

By strobing in another column address with the same open row, we can select another bit of data (again via the column address decoder) to steer to the D_{OUT} pin. In this case, however, we have changed the DRAM to the Write mode (Figure 1.14). This allows us to write, with the same row open via the D_{IN} pin in Figure 1.10, to any column address on the open row. *Later, second-generation DRAMs used the same pins for both data input and output to reduce pin count. These bidirectional pins are labeled DQ.*

The final set of timing signals in Figure 1.14 (the right side) read data out of the DRAM with $R/\overline{W}$ HIGH, change $R/\overline{W}$ to a LOW, and then write to the same location. Again, when $\overline{CAS}$ goes HIGH, D_{OUT} goes back to the Hi-Z state.

The remaining modes of operation are simple modifications of page mode. As seen in Figure 1.15, FPM allows the column address to change while $\overline{CAS}$ is LOW. The speed of the DRAM improves by reducing the delay between $\overline{CAS}$ going LOW and valid data present, or accessed, on D_{OUT} (t_{CAC}). EDO is simply an FPM DRAM that doesn't force D_{OUT} to a Hi-Z state when $\overline{CAS}$ goes HIGH. The data out of the DRAM is thus available

Sec. 1.1 DRAM Types and Operation

for a longer period of time, allowing for faster operation. In general, opening the row is the operation that takes the longest amount of time. Once a row is open, the data sitting on the columnlines can be steered to D_{OUT} at a fast rate. Interestingly, using column access modes has been the primary method of boosting DRAM performance over the years.

The other popular modes of operation in second-generation DRAMs were the static column and nibble modes. Static column mode DRAMs used flow-through latches in the column address path. When a column address was changed externally, with $\overline{CAS}$ LOW, the column address fed directly to the column address decoder. (The address wasn't clocked on the falling edge of $\overline{CAS}$.) This increased the speed of the DRAM by preventing the outputs from going into the Hi-Z state with changes in the column address.

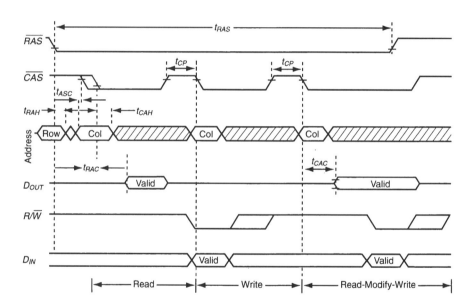

Figure 1.14 Page mode.

Nibble mode DRAMs used an internal presettable address counter so that by strobing $\overline{CAS}$, the column address would change internally. Figure 1.16 illustrates the timing operation for a nibble mode DRAM. The first time $\overline{CAS}$ transitions LOW (first being defined as the first transition after $\overline{RAS}$ goes LOW), the column address is loaded into the counter. If $\overline{RAS}$ is held LOW and $\overline{CAS}$ is toggled, the internal address counter is incremented, and the sequential data appears on the output of the DRAM. The term *nibble mode* comes from limiting the number of $\overline{CAS}$ cycles to four (a nibble).

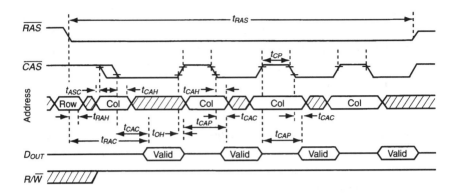

Figure 1.15 Fast page mode.

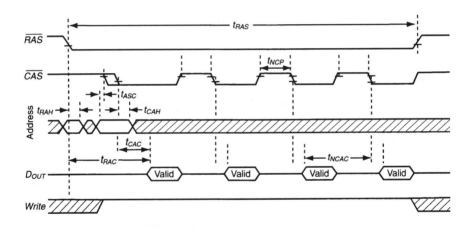

Figure 1.16 Nibble mode.

1.1.3 Synchronous DRAMs (Third Generation)

Synchronous DRAMs (SDRAMs) are made by adding a synchronous interface between the basic core DRAM operation/circuitry of second-generation DRAMs and the control coming from off-chip to make the DRAM operation faster. All commands and operations to and from the DRAM are executed on the rising edge of a master or command clock signal that is common to all SDRAMs and labeled *CLK*. See Figure 1.17 for the pin connections of a 64Mb SDRAM with 16-bit input/output (I/O).

At the time of this writing, SDRAMs operate with a maximum *CLK* frequency in the range of 100–143 MHz. This means that if a 64Mb SDRAM

Sec. 1.1   DRAM Types and Operation

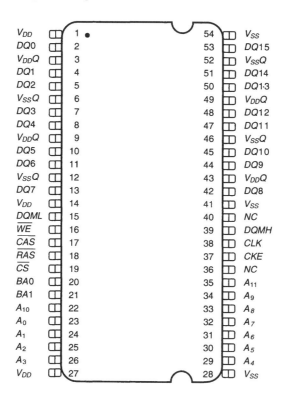

Figure 1.17 Pin connections of a 64Mb SDRAM with 16-bit I/O.

is organized as a x16 part (that is, the input/output word size is 16 bits), the maximum rate at which the words can be written to the part is 200–286 MB/s.

Another variation of the SDRAM is the *double-data-rate SDRAM* (DDR SDRAM, or simply DDR DRAM). The DDR parts register commands and operations on the rising edge of the clock signal while allowing data to be transferred on both the rising and falling edges. A differential input clock signal is used in the DDR DRAM with the labeling of, not surprisingly, *CLK* and $\overline{CLK}$. In addition, the DDR DRAM provides an output data strobe, labeled DQS, synchronized with the output data and the input *CLK*. DQS is used at the controller to strobe in data from a DRAM. The big benefit of using a DDR part is that the data transfer rate can be twice the clock frequency because data can be transferred on both the rising and falling edges of *CLK*. This means that when using a 133 MHz clock, the data written to and read from the DRAM can be transferred at 266M words/s. Using the numbers from the previous paragraph, this means that a 64Mb DDR

SDRAM with an input/output word size of 16 bits will transfer data to and from the memory controller at 400–572 MB/s.

Figure 1.18 shows the block diagram of a 64Mb SDRAM with 16-bit I/O. Note that although *CLK* is now used for transferring data, we still have the second-generation control signals $\overline{CS}$, $\overline{WE}$, $\overline{CAS}$, and $\overline{RAS}$ present on the part. (*CKE* is a clock enable signal which, unless otherwise indicated, is assumed HIGH.) Let's discuss how these control signals are used in an SDRAM by recalling that in a second-generation DRAM, a Write was executed by first driving $\overline{WE}$ and $\overline{CS}$ LOW. Next a row was opened by applying a row address to the part and then driving $\overline{RAS}$ LOW. (The row address is latched on the falling edge of $\overline{RAS}$.) Finally, a column address was applied and latched on the falling edge of $\overline{CAS}$. A short time later, the data applied to the part would be written to the accessed memory location.

For the SDRAM Write, we change the syntax of the descriptions of what's happening in the part. However, the fundamental operation of the DRAM circuitry is the same as that of the second-generation DRAMs. We can list these syntax changes as follows:

1. The memory is segmented into banks. For the 64Mb memory of Figure 1.17 and Figure 1.18, each bank has a size of 16Mbs (organized as 4,096 row addresses [12 bits] x 256 column addresses [8 bits] x 16 bits [16 *DQ* I/O pins]). As discussed earlier, this is nothing more than a simple logic design of the address decoder (although in most practical situations, the banks are also laid out so that they are physically in the same area). The bank selected is determined by the addresses *BA0* and *BA1*.
2. In second-generation DRAMs, we said, "We open a row," as discussed earlier. In SDRAM, we now say, "We activate a row in a bank." We do this by issuing an active command to the part. Issuing an active command is accomplished on the rising edge of *CLK* with a row/bank address applied to the part with $\overline{CS}$ and $\overline{RAS}$ LOW, while $\overline{CAS}$ and $\overline{WE}$ are held HIGH.
3. In second-generation DRAMs, we said, "We write to a location given by a column address," by driving $\overline{CAS}$ LOW with the column address

Sec. 1.1 DRAM Types and Operation

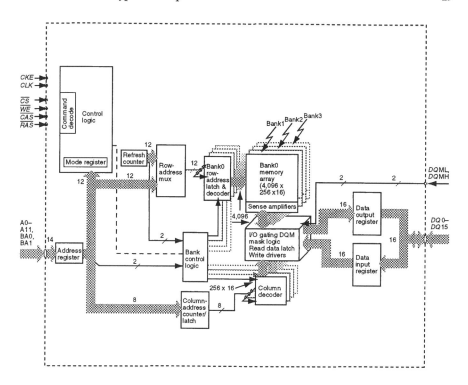

Figure 1.18 Block diagram of a 64Mb SDRAM with 16-bit I/O.

applied to the part and then applying data to the part. In an SDRAM, we write to the part by issuing the Write command to the part. Issuing a Write command is accomplished on the rising edge of *CLK* with a column/bank address applied to the part: $\overline{CS}$, $\overline{CAS}$, and $\overline{WE}$ are held LOW, and $\overline{RAS}$ is held HIGH.

Table 1.1 shows the commands used in an SDRAM. In addition, this table shows how inputs/outputs *(DQs)* can be masked using the *DQ* mask (DQM) inputs. This feature is useful when the DRAM is used in graphics applications.

Table 1.1 SDRAM commands. (Notes: 1)

Name	$\overline{CS}$	$\overline{RAS}$	$\overline{CAS}$	$\overline{WE}$	DQM	ADDR	DQs	Notes
Command inhibit (NOP)	H	X	X	X	X	X	X	—
No operation (NOP)	L	H	H	H	X	X	X	—
Active (select bank and activate row)	L	L	H	H	X	Bank/row	X	3
Read (select bank and column, and start Read burst)	L	H	L	H	L/H[8]	Bank/col	X	4
Write (select bank and column, and start Write burst)	L	H	L	L	L/H[8]	Bank/col	Valid	4
Burst terminate	L	H	H	L	X	X	Active	—
PRECHARGE (deactive row in bank or banks)	L	L	H	L	X	Code	X	5
Auto-Refresh or Self-Refresh (enter Self-Refresh mode)	L	L	L	H	X	X	X	6, 7
Load mode register	L	L	L	L	X	Op-code	X	2
Write Enable/output Enable	—	—	—	—	L	—	Active	8
Write inhibit/output Hi-Z	—	—	—	—	H	—	Hi-Z	8

Notes

1. *CKE* is HIGH for all commands shown except for Self-Refresh.
2. *A0–A11* define the op-code written to the mode register.
3. *A0–A11* provide row address, and *BA0, BA1* determine which bank is made active.
4. *A0–A9* (x4), *A0–A8* (x8), or *A0–A7* (x16) provide column address; *A10* HIGH enables the auto *PRECHARGE* feature (nonpersistent), while *A10* LOW disables the auto *PRECHARGE* feature; *BA0, BA1* determine which bank is being read from or written to.

Sec. 1.1 DRAM Types and Operation

5. *A*10 LOW: *BA*0, *BA*1 determine the bank being precharged. *A*10 HIGH: all banks precharged and *BA*0, *BA*1 are "don't care."
6. This command is Auto-Refresh if *CKE* is HIGH and Self-Refresh if *CKE* is LOW.
7. Internal Refresh counter controls row addressing; all inputs and I/Os are "don't care" except for *CKE*.
8. Activates or deactivates the *DQ*s during Writes (zero-clock delay) and Reads (two-clock delay).

SDRAMs often employ pipelining in the address and data paths to increase operating speed. Pipelining is an effective tool in SDRAM design because it helps disconnect operating frequency and access latency. Without pipelining, a DRAM can only process one access instruction at a time. Essentially, the address is held valid internally until data is fetched from the array and presented to the output buffers. This single instruction mode of operation ties operating frequency and access time (or latency) together. However, with pipelining, additional access instructions can be fed into the SDRAM before prior access instructions have completed, which permits access instructions to be entered at a higher rate than would otherwise be allowed. Hence, pipelining increases operating speed.

Pipeline stages in the data path can also be helpful when synchronizing output data to the system clock. $\overline{CAS}$ latency refers to a parameter used by the SDRAM to synchronize the output data from a Read request with a particular edge of the system clock. A typical Read for an SDRAM with $\overline{CAS}$ latency set to three is shown in Figure 1.19. SDRAMs must be capable of reliably functioning over a range of operating frequencies while maintaining a specified $\overline{CAS}$ latency. This is often accomplished by configuring the pipeline stage to register the output data to a specific clock edge, as determined by the $\overline{CAS}$ latency parameter.

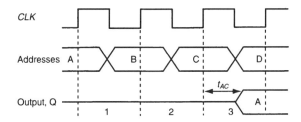

Figure 1.19 SDRAM with a latency of three.

At this point, we should understand the basics of SDRAM operation, but we may be asking, "Why are SDRAMs potentially faster than second-generation DRAMs such as EDO or FPM?" The answer to this question comes from the realization that it's possible to activate a row in one bank and then, while the row is opening, perform an operation in some other bank (such as reading or writing). In addition, one of the banks can be in a *PRECHARGE* mode (the bitlines are driven to $V_{CC}/2$) while accessing one of the other banks and, thus, in effect hiding *PRECHARGE* and allowing data to be continuously written to or read from the SDRAM. (Of course, this depends on which application and memory address locations are used.) We use a mode register, as shown in Figure 1.20, to put the SDRAM into specific modes of operation for programmable operation, including pipelining and burst Reads/Writes of data [2].

1.2 DRAM BASICS

A modern DRAM memory cell or memory bit (mbit), as shown in Figure 1.21, is formed with one transistor and one capacitor, accordingly referred to as a 1T1C cell. The mbit is capable of holding binary information in the form of stored charge on the capacitor. The mbit transistor operates as a switch interposed between the mbit capacitor and the digitline. Assume that the capacitor's common node is biased at $V_{CC}/2$, which we will later show as a reasonable assumption. Storing a logic one in the cell requires a capacitor with a voltage of $+V_{CC}/2$ across it. Therefore, the charge stored in the mbit capacitor is

$$Q = \frac{V_{CC}}{2} \cdot C,$$

where C is the capacitance value in farads. Conversely, storing a logic zero in the cell requires a capacitor with a voltage of $-V_{CC}/2$ across it. Note that the stored charge on the mbit capacitor for a logic zero is

$$Q = \frac{-V_{CC}}{2} \cdot C.$$

The charge is negative with respect to the $V_{CC}/2$ common node voltage in this state. Various leakage paths cause the stored capacitor charge to slowly deplete. To return the stored charge and thereby maintain the stored data state, the cell must be refreshed. The required refreshing operation is what makes DRAM memory dynamic rather than static.

Sec. 1.2 DRAM Basics

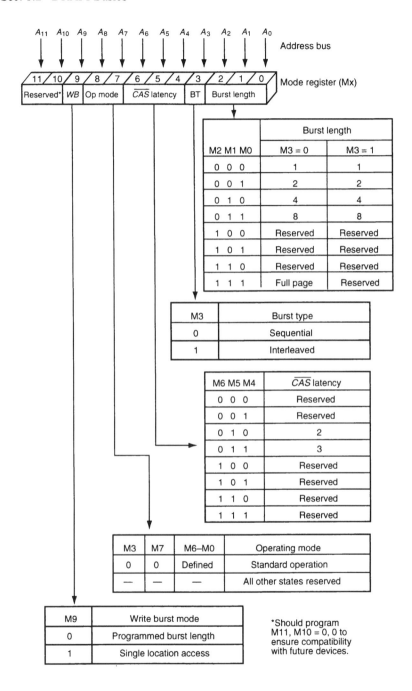

Figure 1.20 Mode register.

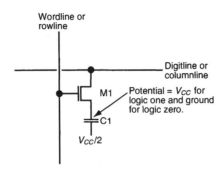

Figure 1.21 1T1C DRAM memory cell.
(Note the rotation of the rowline and columnline.)

The digitline referred to earlier consists of a conductive line connected to a multitude of mbit transistors. The conductive line is generally constructed from either metal or silicide/polycide polysilicon. Because of the quantity of mbits connected to the digitline and its physical length and proximity to other features, the digitline is highly capacitive. For instance, a typical value for digitline capacitance on a 0.35 µm process might be 300fF. Digitline capacitance is an important parameter because it dictates many other aspects of the design. We discuss this further in Section 2.1. For now, we continue describing basic DRAM operation.

The mbit transistor gate terminal is connected to a wordline (rowline). The wordline, which is connected to a multitude of mbits, is actually formed of the same polysilicon as that of the transistor gate. The wordline is physically orthogonal to the digitline. A memory array is formed by tiling a selected quantity of mbits together such that mbits along a given digitline do not share a common wordline and mbits along a common wordline do not share a common digitline. Examples of this are shown in Figures 1.22 and 1.23. In these layouts, mbits are paired to share a common contact to the digitline, which reduces the array size by eliminating duplication.

Sec. 1.2 DRAM Basics

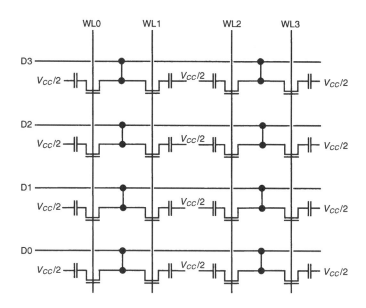

Figure 1.22 Open digitline memory array schematic.

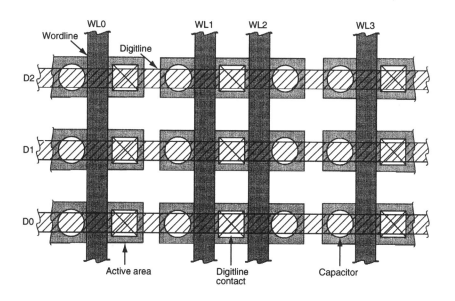

Figure 1.23 Open digitline memory array layout.

1.2.1 Access and Sense Operations

Next, we examine the access and sense operations. We begin by assuming that the cells connected to D1, in Figure 1.24, have logic one levels ($+V_{CC}/2$) stored on them and that the cells connected to D0 have logic zero levels ($-V_{CC}/2$) stored on them. Next, we form a digitline pair by considering two digitlines from adjacent arrays. The digitline pairs, labeled D0/D0* and D1/D1*, are initially equilibrated to $V_{CC}/2$ V. All wordlines are initially at 0 V, ensuring that the mbit transistors are OFF. Prior to a wordline firing, the digitlines are electrically disconnected from the $V_{CC}/2$ bias voltage and allowed to float. They remain at the $V_{CC}/2$ *PRECHARGE* voltage due to their capacitance.

To read mbit1, wordline WL0 changes to a voltage that is at least one transistor V_{TH} above V_{CC}. This voltage level is referred to as V_{CCP} or V_{PP}. To ensure that a full logic one value can be written back into the mbit capacitor, V_{CCP} must remain greater than one V_{TH} above V_{CC}. The mbit capacitor begins to discharge onto the digitline at two different voltage levels depending on the logic level stored in the cell. For a logic one, the capacitor begins to discharge when the wordline voltage exceeds the digitline *PRECHARGE* voltage by V_{TH}. For a logic zero, the capacitor begins to discharge when the wordline voltage exceeds V_{TH}. Because of the finite rise time of the wordline voltage, this difference in turn-on voltage translates into a significant delay when reading ones, as seen in Figure 1.25.

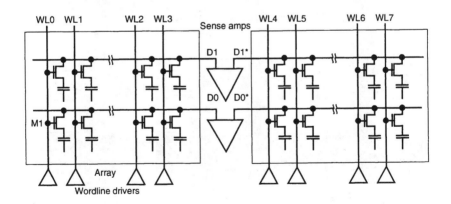

Figure 1.24 Simple array schematic (an open DRAM array).

Sec. 1.2 DRAM Basics

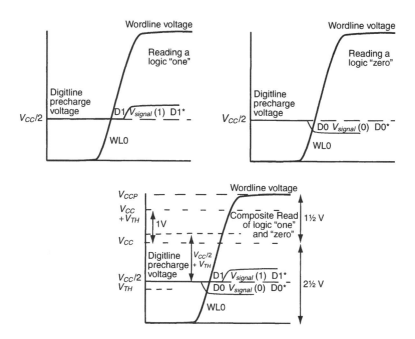

Figure 1.25 Cell access waveforms.

Accessing a DRAM cell results in charge-sharing between the mbit capacitor and the digitline capacitance. This charge-sharing causes the digitline voltage either to increase for a stored logic one or to decrease for a stored logic zero. Ideally, only the digitline connected to the accessed mbit will change. In reality, the other digitline voltage also changes slightly, due to parasitic coupling between digitlines and between the firing wordline and the other digitline. (This is especially true for the folded bitline architecture discussed later.) Nevertheless, a differential voltage develops between the two digitlines. The magnitude of this voltage difference, or signal, is a function of the *mbit capacitance* (C_{mbit}), *digitline capacitance* (C_{digit}), and voltage stored on the cell prior to access (V_{cell}). See Figure 1.26. Accordingly,

$$V_{signal} = V_{cell} \cdot \frac{C_{mbit}}{C_{digit} + C_{mbit}}.$$

A V_{signal} of 235mV is yielded from a design in which V_{cell} = 1.65, C_{mbit} = 50fF, and C_{digit} = 300fF.

After the cell has been accessed, sensing occurs. Sensing is essentially the amplification of the digitline signal or the differential voltage between the digitlines. Sensing is necessary to properly read the cell data and refresh the mbit cells. (The reason for forming a digitline pair now becomes apparent.) Figure 1.27 presents a schematic diagram for a simplified sense amplifier circuit: a cross-coupled NMOS pair and a cross-coupled PMOS pair. The sense amplifiers also appear like a pair of cross-coupled inverters in which *ACT* and *NLAT** provide power and ground. The NMOS pair or Nsense-amp has a common node labeled *NLAT** (for *Nsense-amp latch*).

Similarly, the Psense-amp has a common node labeled *ACT* (for *Active pull-up*). Initially, *NLAT** is biased to $V_{CC}/2$, and *ACT* is biased to V_{SS} or signal ground. Because the digitline pair D1 and D1* are both initially at $V_{CC}/2$, the Nsense-amp transistors are both OFF. Similarly, both Psense-amp transistors are OFF. Again, when the mbit is accessed, a signal develops across the digitline pair. While one digitline contains charge from the cell access, the other digitline does not but serves as a reference for the Sensing operation. The sense amplifiers are generally fired sequentially: the Nsense-amp first, then the Psense-amp. Although designs vary at this point, the higher drive of NMOS transistors and better V_{TH} matching offer better sensing characteristics by Nsense-amps and thus lower error probability compared to Psense-amps.

Waveforms for the Sensing operation are shown in Figure 1.28. The Nsense-amp is fired by bringing *NLAT** (N sense-amp latch) toward ground. As the voltage difference between *NLAT** and the digitlines (D1 and D1* in Figure 1.27) approaches V_{TH}, the NMOS transistor whose gate is connected to the higher voltage digitline begins to conduct. This conduction occurs first in the subthreshold and then in the saturation region as the gate-to-source voltage exceeds V_{TH} and causes the low-voltage digitline to discharge toward the *NLAT** voltage. Ultimately, *NLAT** will reach ground and the digitline will be brought to ground potential. Note that the other NMOS transistor will not conduct: its gate voltage is derived from the low-

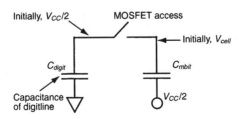

Figure 1.26 DRAM charge-sharing.

Sec. 1.2 DRAM Basics

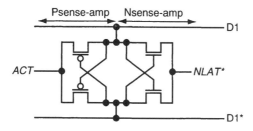

Figure 1.27 Sense amplifier schematic.

voltage digitline, which is being discharged toward ground. In reality, parasitic coupling between digitlines and limited subthreshold conduction by the second transistor result in voltage reduction on the high digitline.

Sometime after the Nsense-amp fires, *ACT* will be brought toward V_{CC} to activate the Psense-amp, which operates in a complementary fashion to the Nsense-amp. With the low-voltage digitline approaching ground, there is a strong signal to drive the appropriate PMOS transistor into conduction. This conduction, again moving from subthreshold to saturation, charges the high-voltage digitline toward *ACT*, ultimately reaching V_{CC}. Because the mbit transistor remains ON, the mbit capacitor is refreshed during the Sensing operation. The voltage, and hence charge, which the mbit capacitor held prior to accessing, is restored to a full level: V_{CC} for a logic one and ground for a logic zero. It should be apparent now why the minimum wordline voltage is a V_{TH} above V_{CC}. If V_{CCP} were anything less, a full V_{CC} level could not be written back into the mbit capacitor. The mbit transistor source voltage V_{source} cannot be greater than $V_{gate} - V_{TH}$ because this would turn OFF the transistor.

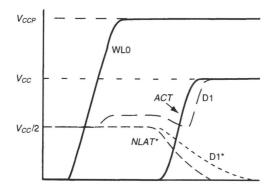

Figure 1.28 Sensing operation waveforms.

1.2.2 Write Operation

A Write operation is similar to a Sensing and Restore operation except that a separate Write driver circuit determines the data that is placed into the cell. The Write driver circuit is generally a tristate inverter connected to the digitlines through a second pair of pass transistors, as shown in Figure 1.29. These pass transistors are referred to as I/O transistors. The gate terminals of the I/O transistors are connected to a common *column select* (*CSEL*) signal. The *CSEL* signal is decoded from the column address to select which pair (or multiple pairs) of digitlines is routed to the output pad or, in this case, the Write driver.

In most current DRAM designs, the Write driver simply overdrives the sense amplifiers, which remain ON during the Write operation. After the new data is written into the sense amplifiers, the amplifiers finish the Write cycle by restoring the digitlines to full rail-to-rail voltages. An example is shown in Figure 1.30 in which D1 is initially HIGH after the Sensing operation and LOW after the writing operation. A Write operation usually involves only 2–4 mbits within an array of mbits because a single *CSEL* line is generally connected to only four pairs of I/O transistors. The remaining digitlines are accessed through additional *CSEL* lines that correspond to different column address locations.

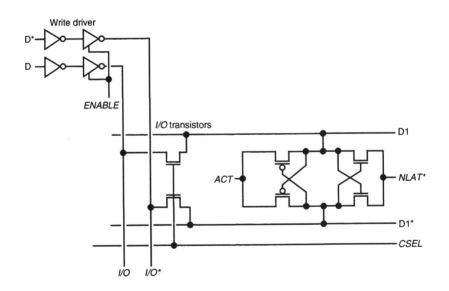

Figure 1.29 Sense amplifier schematic with I/O devices.

Sec. 1.2 DRAM Basics   31

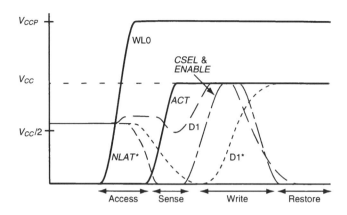

Figure 1.30 Write operation waveforms.

1.2.3 Opening a Row (Summary)

Opening a row of mbits in a DRAM array is a fundamental operation for both reading and writing to the DRAM array. Sometimes the chain of events from a circuit designer's point of view, which lead to an open row, is called the $\overline{RAS}$ timing chain. We summarize the $\overline{RAS}$ timing chain of events below, assuming that for a second-generation DRAM both $\overline{RAS}$ and $\overline{CAS}$ are HIGH. (It's trivial to extend our discussion to third-generation DRAMs where $\overline{RAS}$ and $\overline{CAS}$ are effectively generated from the control logic.)

1. Initially, both $\overline{RAS}$ and $\overline{CAS}$ are HIGH. All bitlines in the DRAM are driven to $V_{CC}/2$, while all wordlines are at 0 V. This ensures that all of the mbit's access transistors in the DRAM are OFF.

2. A valid row address is applied to the DRAM and $\overline{RAS}$ goes LOW. While the row address is being latched, on the falling edge of $\overline{RAS}$, and decoded, the bitlines are disconnected from the $V_{CC}/2$ bias and allowed to float. The bitlines at this point are charged to $V_{CC}/2$, and they can be thought of as capacitors.

3. The row address is decoded and applied to the wordline drivers. This forces only one rowline in at least one memory array to V_{CCP}. Driving the wordline to V_{CCP} turns ON the mbits attached to this rowline and causes charge-sharing between the mbit capacitance and the capacitance of the corresponding bitline. The result is a small perturbation (upwards for a logic one and downwards for a logic zero) in the bitline voltages.

4. The next operation is Sensing, which has two purposes: (a) to determine if a logic one or zero was written to the cell and (b) to refresh the contents of the cell by restoring a full logic zero (0 V) or one *(V_{CC})* to the capacitor. Following the wordlines going HIGH, the Nsense-amp is fired by driving, via an n-channel MOSFET, *NLAT** to ground. The inputs to the sense amplifier are two bitlines: the bitline we are sensing and the bitline that is not active (a bitline that is still charged to $V_{CC}/2$—an inactive bitline). Pulling *NLAT** to ground results in one of the bitlines going to ground. Next, the *ACT* signal is pulled up to V_{CC}, driving the other bitline to V_{CC}. Some important notes:

   (a) It doesn't matter if a logic one or logic zero was sensed because the inactive and active bitlines are pulled in opposite directions.
   (b) The contents of the active cell, after opening a row, are restored to full voltage levels (either 0 V or V_{CC}). The entire DRAM can be refreshed by opening each row.

Now that the row is open, we can write to or read from the DRAM. In either case, it is a simple matter of steering data to or from the active array(s) using the column decoder. When writing to the array, buffers set the new logic voltage levels on the bitlines. The row is still open because the wordline remains HIGH. (The row stays open as long as $\overline{RAS}$ is LOW.)

When reading data out of the DRAM, the values sitting on the bitlines are transmitted to the output buffers via the I/O MOSFETs. To increase the speed of the reading operation, this data, in most situations, is transmitted to the output buffer (sometimes called a *DQ* buffer) either through a helper flip-flop or another sense amplifier.

A note is in order here regarding the word size stored in or read out of the memory array. We may have 512 active bitlines when a single rowline in an array goes HIGH (keeping in mind once again that only one wordline in an array can go HIGH at any given time). This literally means that we could have a word size of 512 bits from the active array. The inherent wide word size has led to the push, at the time of this writing, of embedding DRAM with a processor (for example, graphics or data). The wide word size and the fact that the word doesn't have to be transmitted off-chip can result in lower-power, higher-speed systems. (Because the memory and processor don't need to communicate off-chip, there is no need for power-hungry, high-speed buffers.)

Sec. 1.2 DRAM Basics

1.2.4 Open/Folded DRAM Array Architectures

Throughout the book, we make a distinction between the open array architecture as shown in Figures 1.22 and 1.24 and the folded DRAM array used in modern DRAMs and seen in Figure 1.31. At the cost of increased layout area, folded arrays increase noise immunity by moving sense amp inputs next to each other. These sense amp inputs come directly from the DRAM array. The term *folded* comes from taking the DRAM arrays seen in Figure 1.24 and folding them together to form the topology seen in Figure 1.31.

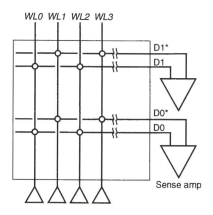

Figure 1.31 A folded DRAM array.

REFERENCES

[1] R. J. Baker, H. W. Li, and D. E. Boyce, *CMOS: Circuit Design, Layout, and Simulation.* Piscataway, NJ: IEEE Press, 1998.

[2] Micron Technology, Inc., Synchronous DRAM Data Sheet, 1999.

FOR FURTHER REFERENCE

See the Appendix for additional readings and references.

Chapter 2

The DRAM Array

This chapter begins a more detailed examination of standard DRAM array elements. This examination is necessary for a clear understanding of fundamental DRAM elements and how they are used in memory block construction. A common point of reference is required before considering the analysis of competing array architectures. Included in this chapter is a detailed discussion of mbits, array configurations, sense amplifier elements, and row decoder elements.

2.1 THE MBIT CELL

The primary advantage of DRAM over other types of memory technology is low cost. This advantage arises from the simplicity and scaling characteristics of its 1T1C memory cell [1]. Although the DRAM mbit is simple conceptually, its actual design and implementation are highly complex. Therefore, successful, cost-effective DRAM designs require a tremendous amount of process technology.

Figure 2.1 presents the layout of a modern buried capacitor DRAM mbit pair. (*Buried* means that the capacitor is below the digitline.) This type of mbit is also referred to as a *bitline over capacitor* (BOC) cell. Because sharing a contact significantly reduces overall cell size, DRAM mbits are constructed in pairs. In this way, a digitline contact can be shared. The mbits comprise an active area rectangle (in this case, an n+ active area), a pair of polysilicon wordlines, a single digitline contact, a metal or polysilicon digitline, and a pair of cell capacitors formed with an oxide-nitride-oxide dielectric between two layers of polysilicon. For most processes, the wordline polysilicon is silicided to reduce sheet resistance, permitting longer wordline segments without reducing speed. The mbit layout, as shown in Figure

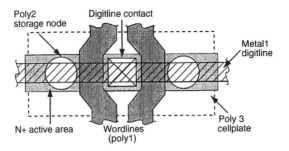

Figure 2.1 Mbit pair layout.

2.1, is essentially under the control of process engineers, for every aspect of the mbit must meet stringent performance and yield criteria.

A small array of mbits appears in Figure 2.2. This figure is useful to illustrate several features of the mbit. First, note that the digitline pitch (width plus space) dictates the active area pitch and the capacitor pitch. Process engineers adjust the active area width and the field oxide width to maximize transistor drive and minimize transistor-to-transistor leakage. Field oxide technology greatly impacts this balance. A thicker field oxide or a shallower junction depth affords a wider transistor active area. Second, the wordline pitch (width plus space) dictates the space available for the digitline contact, transistor length, active area, field poly width, and capacitor length. Optimization of each of these features by process engineers is necessary to maximize capacitance, minimize leakage, and maximize yield. Contact technology, subthreshold transistor characteristics, photolithography, and etch and film technology dictate the overall design.

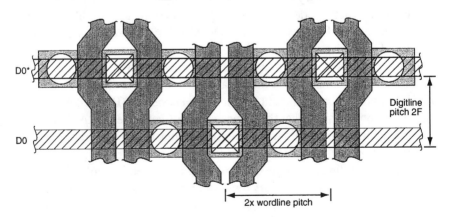

Figure 2.2 Layout to show array pitch.

Sec. 2.1 The Mbit Cell

At this point in the discussion, it is appropriate to introduce the concept of feature size and how it relates to cell size. The mbit shown in Figures 2.1 and 2.2 is by definition an eight-square feature cell (8F²) [2][3]. The intended definition of feature (F) in this case is minimum realizable process dimension but in fact equates to a dimension that is one-half the wordline (row) or digitline (column) pitch. A 0.25 µm process having wordline and digitline pitches of 0.6 µm yields an mbit size that is

$$(8) \cdot (0.3 \ \mu m)^2 = 0.72 \ \mu m^2.$$

It is easier to explain the 8F² designation with the aid of Figure 2.3. An imaginary box drawn around the mbit defines the cell's outer boundary. Along the x-axis, this box includes one-half digitline contact feature, one wordline feature, one capacitor feature, one field poly feature, and one-half poly space feature for a total of four features. Along the y-axis, this box contains two one-half field oxide features and one active area feature for a total of two features. The area of the mbit is therefore

$$4F \cdot 2F = 8F^2.$$

The folded array architecture, as shown in Figure 2.2, always produces an 8F² mbit. This results from the fact that each wordline connects or forms a crosspoint with an mbit transistor on every other digitline and must pass around mbit transistors as field poly on the remaining digitlines. The field poly in each mbit cell adds two square features to what would have been a 6F² cell. Although the folded array yields a cell that is 25% larger than other array architectures, it also produces superior signal-to-noise performance, especially when combined with some form of digitline twisting [4]. Superior low-noise performance has made folded array architecture the architecture of choice since the 64kbit generation [5].

A folded array is schematically depicted in Figure 2.4. Sense amplifier circuits placed at the edge of each array connect to both true and complement digitlines *(D* and *D*)* coming from a single array. Optional digitline pair twisting in one or more positions reduces and balances the coupling to adjacent digitline pairs and improves overall signal-to-noise characteristics [4]. Figure 2.5 shows the variety of twisting schemes used throughout the DRAM industry [6].

Ideally, a twisting scheme equalizes the coupling terms from each digitline to all other digitlines, both true and complement. If implemented properly, the noise terms cancel or at least produce only common-mode noise to which the differential sense amplifier is more immune.

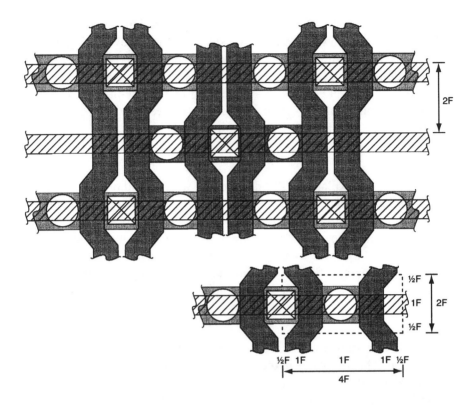

Figure 2.3 Layout to show 8F² derivation.

Each digitline twist region consumes valuable silicon area. Thus, design engineers resort to the simplest and most efficient twisting scheme to get the job done. Because the coupling between adjacent metal lines is inversely proportional to the line spacing, the signal-to-noise problem gets increasingly worse as DRAMs scale to smaller and smaller dimensions. Hence, the industry trend is toward use of more complex twisting schemes on succeeding generations [6][7].

An alternative to the folded array architecture, popular prior to the 64kbit generation [1], was the open digitline architecture. Seen schematically in Figure 2.6, this architecture also features the sense amplifier circuits between two sets of arrays [8]. Unlike the folded array, however, true and complement digitlines *(D* and *D*)* connected to each sense amplifier pair come from separate arrays [9]. This arrangement precludes using digitline twisting to improve signal-to-noise performance, which is the prevalent reason why the industry switched to folded arrays. Note that unlike the folded array architecture, each wordline in an open digitline architecture connects to mbit transistors on every digitline, creating crosspoint-style arrays.

Sec. 2.1 The Mbit Cell

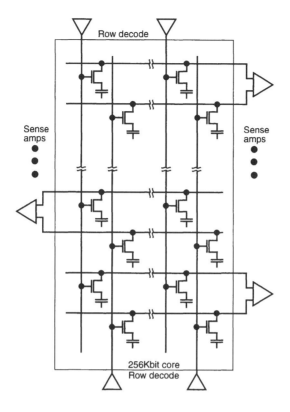

Figure 2.4 Folded digitline array schematic.

This feature permits a 25% reduction in mbit size to only $6F^2$ because the wordlines do not have to pass alternate mbits as field poly. The layout for an array of standard $6F^2$ mbit pairs is shown in Figure 2.7 [2]. A box is drawn around one of the mbits to show the $6F^2$ cell boundary. Again, two mbits share a common digitline contact to improve layout efficiency. Unfortunately, most manufacturers have found that the signal-to-noise problems of open digitline architecture outweigh the benefits derived from reduced array size [8].

Digitline capacitive components, contributed by each mbit, include junction capacitance, digitline-to-cellplate (poly3), digitline-to-wordline, digitline-to-digitline, digitline-to-substrate, and, in some cases, digitline-to-storage cell (poly2) capacitance. Therefore, each mbit connected to the digitline adds a specific amount of capacitance to the digitline. Most modern DRAM designs have no more than 256 mbits connected to a digitline segment.

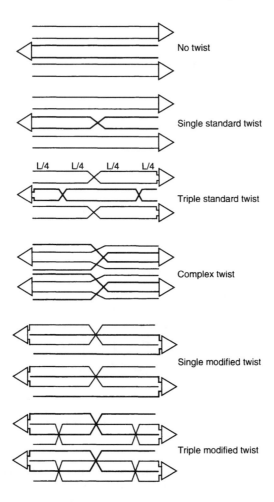

Figure 2.5 Digitline twist schemes.

Two factors dictate this quantity. First, for a given cell size, as determined by row and column pitches, a maximum storage capacitance can be achieved without resorting to exotic processes or excessive cell height. For processes in which the digitline is above the storage capacitor (buried capacitor), contact technology determines the maximum allowable cell height. This fixes the volume available (cell area multiplied by cell height) in which to build the storage capacitor. Second, as the digitline capacitance increases, the power associated with charging and discharging this capacitance during Read and Write operations increases. Any given wordline

Sec. 2.1 The Mbit Cell

essentially accesses (crosses) all of the columns within a DRAM. For a 256-Meg DRAM, each wordline crosses 16,384 columns. With a multiplier such as that, it is easy to appreciate why limits to digitline capacitance are necessary to keep power dissipation low.

Figure 2.8 presents a process cross section for the buried capacitor mbit depicted in Figure 2.2, and Figure 2.9 shows a SEM image of the buried capacitor mbit. This type of mbit, employing a buried capacitor structure, places the digitline physically above the storage capacitor [10]. The digitline is constructed from either metal or polycide, while the digitline contact is formed using a metal or polysilicon plug technology. The mbit capacitor is formed with polysilicon (poly2) as the bottom plate, an oxide-nitride-oxide (ONO) dielectric, and a sheet of polysilicon (poly3). This top sheet of polysilicon becomes a common node shared by all mbit capacitors. The capacitor shape can be simple, such as a rectangle, or complex, such as concentric cylinders or stacked discs. The most complex capacitor structures are the topic of many DRAM process papers [11][12][13].

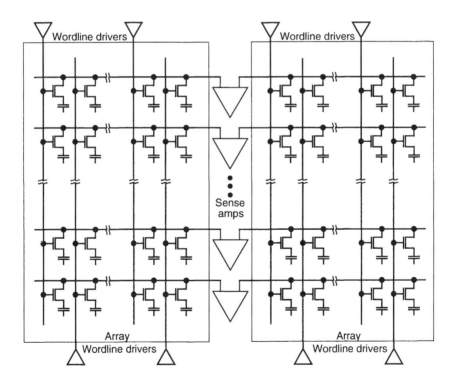

Figure 2.6 Open digitline array schematic.

The ONO dielectric undergoes optimization to achieve maximum capacitance with minimum leakage. It must also tolerate the maximum DRAM operating voltage without breaking down. For this reason, the cellplate (poly3) is normally biased at $+V_{CC}/2$ V, ensuring that the dielectric has no more than $V_{CC}/2$ V across it for either stored logic state, a logic one at $+V_{CC}/2$ V or a logic zero at $-V_{CC}/2$ V.

Two other basic mbit configurations are used in the DRAM industry. The first, shown in Figures 2.10, 2.11, and 2.12, is referred to as a *buried digitline* or *capacitor over bitline* (COB) cell [14][15]. The digitline in this cell is almost always formed of polysilicon rather than of metal.

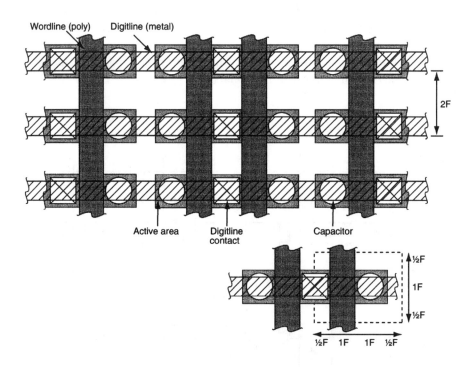

Figure 2.7 Open digitline array layout.
(Feature size (F) is equal to one-half digitline pitch.)

Sec. 2.1 The Mbit Cell

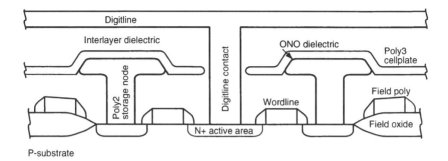

Figure 2.8 Buried capacitor cell process cross section.

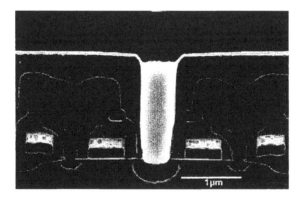

Figure 2.9 Buried capacitor cell process SEM image.

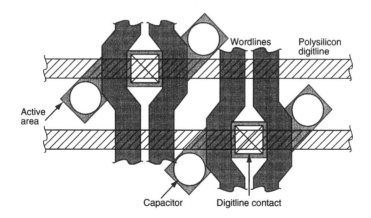

Figure 2.10 Buried digitline mbit cell layout.

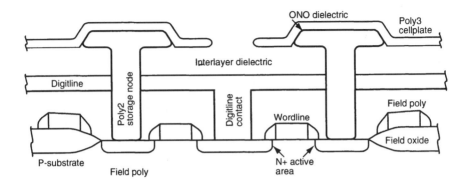

Figure 2.11 Buried digitline mbit process cross section.

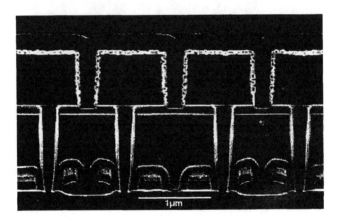

Figure 2.12 Buried digitline mbit process SEM image.

As viewed from the top, the active area is normally bent or angled to accommodate the storage capacitor contact that must drop between digitlines. An advantage of the buried digitline cell over the buried capacitor cell of Figure 2.8 is that its digitline is physically very close to the silicon surface, making digitline contacts much easier to produce. The angled active area, however, reduces the effective active area pitch, constraining the isolation process even further. In buried digitline cells, it is also very difficult to form the capacitor contact. Because the digitline is at or near minimum pitch for the process, insertion of a contact between digitlines can be difficult.

Sec. 2.1 The Mbit Cell   45

Figures 2.13 and 2.14 present a process cross section of the third type of mbit used in the construction of DRAMs. Using trench storage capacitors, this cell is accordingly called a *trench cell* [12][13]. Trench capacitors are formed in the silicon substrate, rather than above the substrate, after etching deep holes into the wafer. The storage node is a doped polysilicon plug, which is deposited in the hole following growth or deposition of the capacitor dielectric. Contact between the storage node plug and the transistor drain is usually made through a poly strap.

With most trench capacitor designs, the substrate serves as the common-node connection to the capacitors, preventing the use of $+V_{CC}/2$ bias and thinner dielectrics. The substrate is heavily doped around the capacitor to reduce resistance and improve the capacitor's CV characteristics. A real advantage of the trench cell is that the capacitance can be increased by merely etching a deeper hole into the substrate [16].

Furthermore, the capacitor does not add stack height to the design, greatly simplifying contact technology. The disadvantage of trench capacitor technology is the difficulty associated with reliably building capacitors in deep silicon holes and connecting the trench capacitor to the transistor drain terminal.

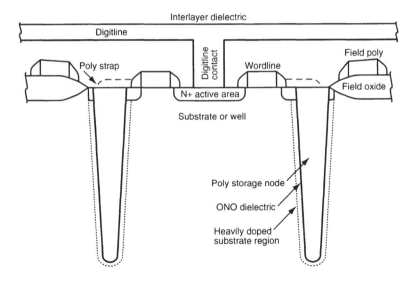

Figure 2.13 Trench capacitor mbit process cross section.

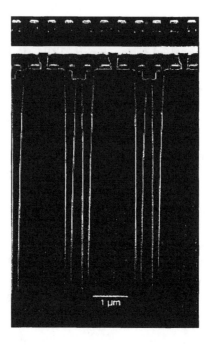

Figure 2.14 Trench capacitor mbit process SEM image.

2.2 THE SENSE AMP

The term *sense amplifier* actually refers to a collection of circuit elements that pitch up to the digitlines of a DRAM array. This collection most generally includes isolation transistors, devices for digitline equilibration and bias, one or more Nsense-amplifiers, one or more Psense-amplifiers, and devices connecting selected digitlines to *I/O* signal lines. All of the circuits along with the wordline driver circuits, as discussed in Section 2.3, are called *pitch cells*. This designation comes from the requirement that the physical layout for these circuits is constrained by the digitline and wordline pitches of an array of mbits. For example, the sense amplifiers for a specific digitline pair (column) are generally laid out within the space of four digitlines. With one sense amplifier for every four digitlines, this is commonly referred to as quarter pitch or four pitch.

2.2.1 Equilibration and Bias Circuits

The first elements analyzed are the equilibration and bias circuits. As we can recall from the discussions of DRAM operation in Section 1.1, the digitlines start at $V_{CC}/2$ V prior to cell Access and Sensing [17]. It is vitally

Sec. 2.2 The Sense Amp

important to the Sensing operation that both digitlines, which form a column pair, are of the same voltage before the wordline is fired. Any offset voltage appearing between the pair directly reduces the effective signal produced during the Access operation [5]. Equilibration of the digitlines is accomplished with one or more NMOS transistors connected between the digitline conductors. NMOS is used because of its higher drive capability and the resulting faster equilibration.

An equilibration transistor, together with bias transistors, is shown in Figure 2.15. The gate terminal is connected to a signal called *equilibrate (EQ)*. *EQ* is held to V_{CC} whenever the external row address strobe signal $\overline{RAS}$ is HIGH. This indicates an inactive or *PRECHARGE* state for the DRAM. After $\overline{RAS}$ has fallen, *EQ* transitions LOW, turning the equilibration transistor OFF just prior to any wordline firing. *EQ* will again transition HIGH at the end of a $\overline{RAS}$ cycle to force equilibration of the digitlines. The equilibration transistor is sized large enough to ensure rapid equilibration of the digitlines to prepare the part for a subsequent access.

As shown in Figure 2.15, two more NMOS transistors accompany the *EQ* transistor to provide a bias level of $V_{CC}/2$ V. These devices operate in conjunction with equilibration to ensure that the digitline pair remains at the prescribed voltage for Sensing. Normally, digitlines that are at V_{CC} and ground equilibrate to $V_{CC}/2$ V [5]. The bias devices ensure that this occurs and also that the digitlines remain at $V_{CC}/2$, despite leakage paths that would otherwise discharge them. Again, for the same reasons as for the equilibration transistor, NMOS transistors are used. Most often, the bias and equilibration transistors are integrated to reduce their overall size. $V_{CC}/2$ V *PRECHARGE* is used on most modern DRAMs because it reduces power consumption and Read-Write times and improves Sensing operations. Power consumption is reduced because a $V_{CC}/2$ *PRECHARGE* voltage can be obtained by equilibrating the digitlines (which are at V_{CC} and ground, respectively) at the end of each cycle.

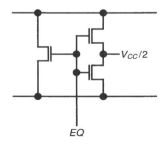

Figure 2.15 Equilibration schematic.

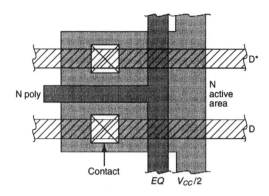

Figure 2.16 Equilibration and bias circuit layout.

The charge-sharing between the digitlines produces $V_{CC}/2$ without additional I_{CC} current. The IBM® PMOS 16-Meg mbit DRAM designs are exceptions: they equilibrate and bias the digitlines to V_{CC} [18]. Because the wordlines and digitlines are both at V_{CC} when the part is inactive, row-to-column shorts that exist do not contribute to increased standby current. $V_{CC}/2$ *PRECHARGE* DRAMs, on the other hand, suffer higher standby current with row-to-column shorts because the wordlines and digitlines are at different potentials when the part is inactive. A layout for the equilibration and bias circuits is shown in Figure 2.16.

2.2.2 Isolation Devices

Isolation devices are important to the sense amplifier circuits. These devices are NMOS transistors placed between the array digitlines and the sense amplifiers (see Figure 2.19). Isolation transistors are physically located on both ends of the sense amplifier layout. In quarter-pitch sense amplifier designs, there is one isolation transistor for every two digitlines. Although this is twice the active area width and space of an array, it nevertheless sets the limit for isolation processing in the pitch cells.

The isolation devices provide two functions. First, if the sense amps are positioned between and connected to two arrays, they electrically isolate one of the two arrays. This is necessary whenever a wordline fires in one

Sec. 2.2 The Sense Amp 49

array because isolation of the second array reduces the digitline capacitance driven by the sense amplifiers, thus speeding Read-Write times, reducing power consumption, and extending Refresh for the isolated array. Second, the isolation devices provide resistance between the sense amplifier and the digitlines. This resistance stabilizes the sense amplifiers and speeds up the Sensing operation by somewhat isolating the highly capacitive digitlines from the low-capacitance sense nodes [19]. Capacitance of the sense nodes between isolation transistors is generally less than 15fF, permitting the sense amplifier to latch much faster than if it were solidly connected to the digitlines. The isolation transistors slow Write-Back to the mbits, but this is far less of a problem than initial Sensing.

2.2.3 Input/Output Transistors

The input/output (I/O) transistors allow data to be read from and written to specific digitline pairs. A single I/O transistor is connected to each sense node as shown in Figure 2.17. The outputs of each I/O transistor are connected to *I/O* signal pairs. Commonly, there are two pairs of *I/O* signal lines, which permit four I/O transistors to share a single *column select (CSEL)* control signal. DRAM designs employing two or more metal layers run the column select lines across the arrays in either Metal2 or Metal3. Each column select can activate four I/O transistors on each side of an array to connect four digitline pairs (columns) to peripheral data path circuits. The I/O transistors must be sized carefully to ensure that instability is not introduced into the sense amplifiers by the I/O bias voltage or remnant voltages on the I/O lines.

Although designs vary significantly as to the numerical ratio, I/O transistors are designed to be two to eight times smaller than the Nsense-amplifier transistors. This is sometimes referred to as *beta ratio*. A beta ratio between five and eight is considered standard; however, it can only be verified with silicon. Simulations fail to adequately predict sense amplifier instability, although theory would predict better stability with higher beta ratio and better Write times with lower beta ratio. During Write, the sense amplifier remains ON and must be overdriven by the Write driver (see Section 1.2.2).

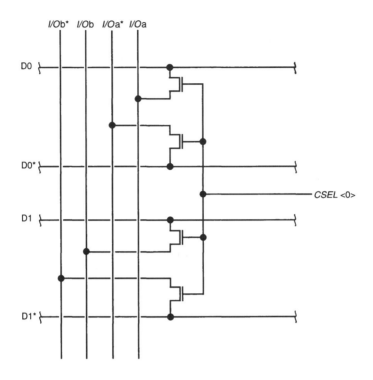

Figure 2.17 I/O transistors.

2.2.4 Nsense- and Psense-Amplifiers

The fundamental elements of any sense amplifier block are the Nsense-amplifier and the Psense-amplifier. These amplifiers, as previously discussed, work together to detect the access signal voltage and drive the digit-lines, accordingly to V_{CC} and ground. The Nsense-amplifier depicted in Figure 2.18 consists of cross-coupled NMOS transistors. The Nsense-amplifier drives the low-potential digitline to ground. Similarly, the Psense-amplifier consists of cross-coupled PMOS transistors and drives the high-potential digitline to V_{CC}.

The sense amplifiers are carefully designed to guarantee correct detection and amplification of the small signal voltage produced during cell access (less than 200mV) [5]. Matching of transistor V_{TH}, transconductance, and junction capacitance within close tolerances helps ensure reliable sense amplifier operation. Ultimately, the layout dictates the overall balance and performance of the sense amplifier block. As a result, a tremendous amount of time is spent ensuring that the sense amplifier layout is optimum. Sym-

Sec. 2.2 The Sense Amp 51

metry and exact duplication of elements are critical to a successful design. This includes balanced coupling to all sources of noise, such as *I/O* lines and latch signals *(NLAT* and ACT)*. Balance is especially critical for layout residing inside the isolation transistors. And because the sense node capacitance is very low, it is more sensitive to noise and circuit imbalances.

While the majority of DRAM designs latch the digitlines to V_{CC} and ground, a growing number of designs are beginning to reduce these levels. Various technical papers report improved Refresh times and lower power dissipation through reductions in latch voltages [20][21]. At first, this appears contradictory: writing a smaller charge into the memory cell would require refreshing the cell more often. However, by maintaining lower *drain-to-source voltages (V_{DS})* and negative *gate-to-source voltages (V_{GS})* across nonaccessed mbit transistors, substantially lower subthreshold leakage and longer Refresh times can be realized, despite the smaller stored charge. An important concern that we are not mentioning here is the leakage resulting from defects in the silicon crystal structure. These defects can result in an increase in the drain/substrate diode saturation current and can practically limit the Refresh times.

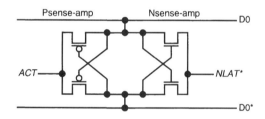

Figure 2.18 Basic sense amplifier block.

Most designs that implement reduced latch voltages generally raise the Nsense-amplifier latch voltage without lowering the Psense-amplifier latch voltage. Designated as boosted sense ground designs, they write data into each mbit using full V_{CC} for a logic one and boosted ground for a logic zero. The sense ground level is generally a few hundred millivolts above true ground. In standard DRAMs, which drive digitlines fully to ground, the V_{GS} of nonaccessed mbits becomes zero when the digitlines are latched. This results in high subthreshold leakage for a stored one level because full V_{CC} exists across the mbit transistor while the V_{GS} is held to zero. Stored zero levels do not suffer from prolonged subthreshold leakage: any amount of cell leakage produces a negative V_{GS} for the transistor. The net effect is that a stored one level leaks away much faster than a stored zero level. One's level retention, therefore, establishes the maximum Refresh period for most DRAM designs. Boosted sense ground extends Refresh by reducing sub-threshold leakage for stored ones. This is accomplished by guaranteeing negative gate-to-source bias on nonaccessed mbit transistors. The benefit of extended Refresh from these designs is somewhat diminished, though, by the added complexity of generating boosted ground levels and the problem of digitlines that no longer equilibrate at $V_{CC}/2$ V.

2.2.5 Rate of Activation

The rate at which the sense amplifiers are activated has been the subject of some debate. A variety of designs use multistage circuits to control the rate at which *NLAT** fires. Especially prevalent with boosted sense ground designs are two-stage circuits that initially drive *NLAT** quickly toward true ground to speed sensing and then bring *NLAT** to the boosted ground level to reduce cell leakage. An alternative to this approach, again using two-stage drivers, drives *NLAT** to ground, slowly at first to limit current and digitline disturbances. This is followed by a second phase in which *NLAT** is driven more strongly toward ground to complete the Sensing operation. This phase usually occurs in conjunction with *ACT* activation. Although these two designs have contrary operation, each meets specific performance objectives: trading off noise and speed.

2.2.6 Configurations

Figure 2.19 shows a sense amplifier block commonly used in double- or triple-metal designs. It features two Psense-amplifiers outside the isolation transistors, a pair of *EQ*/bias *(EQ*b*)* devices, a single Nsense-amplifier, and a single I/O transistor for each digitline. Because only half of the sense amplifiers for each array are on one side, this design is quarter pitch, as are the designs in Figures 2.20 and 2.21. Placement of the Psense-amplifiers outside the isolation devices is necessary because a full one level *(V_{CC})* cannot pass through unless the gate terminal of the *ISO* transistors is driven above V_{CC}. *EQ*/bias transistors are placed outside of the *ISO* devices to permit continued equilibration of digitlines in arrays that are isolated. The I/O transistor gate terminals are connected to a common *CSEL* signal for four adjacent digitlines. Each of the four I/O transistors is tied to a separate I/O bus. This sense amplifier, though simple to implement, is somewhat larger than other designs due to the presence of two Psense-amplifiers.

Sec. 2.2 The Sense Amp 53

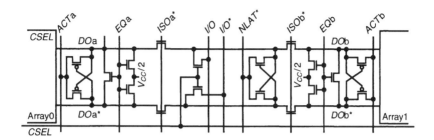

Figure 2.19 Standard sense amplifier block.

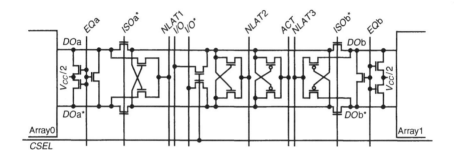

Figure 2.20 Complex sense amplifier block.

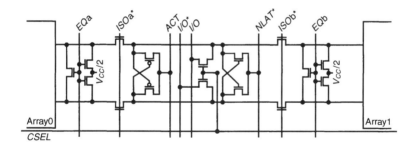

Figure 2.21 Reduced sense amplifier block.

Figure 2.20 shows a second, more complicated style of sense amplifier block. This design employs a single Psense-amplifier and three sets of Nsense-amplifiers. Because the Psense-amplifier is within the isolation transistors, the isolation devices must be the NMOS depletion type, the PMOS enhancement type, or the NMOS enhancement type with boosted gate drive to permit writing a full logic one into the array mbits. The triple Nsense-amplifier is suggestive of PMOS isolation transistors; it prevents full zero levels to be written unless the Nsense-amplifiers are placed adjacent to the arrays. In this more complicated style of sense amplifier block, using three Nsense-amplifiers guarantees faster sensing and higher stability than a similar design using only two Nsense-amplifiers. The inside Nsense-amplifier is fired before the outside Nsense-amplifiers. However, this design will not yield a minimum layout, an objective that must be traded off against performance needs.

The sense amplifier block of Figure 2.21 can be considered a reduced configuration. This design has only one Nsense-amp and one Psense-amp, both of which are placed within the isolation transistors. To write full logic levels, either the isolation transistors must be depletion mode devices or the gate voltage must be boosted above V_{CC} by at least one V_{TH}. This design still uses a pair of *EQ*/bias circuits to maintain equilibration on isolated arrays.

Only a handful of designs operates with a single *EQ*/bias circuit inside the isolation devices, as shown in Figure 2.22. Historically, DRAM engineers tended to shy away from designs that permitted digitlines to float for extended periods of time. However, as of this writing, at least three manufacturers in volume production have designs using this scheme.

A sense amplifier design for single-metal DRAMs is shown in Figure 2.23. Prevalent on 1-Meg and 4-Meg designs, single-metal processes conceded to multi-metal processes at the 16-Meg generation. Unlike the sense amplifiers shown in Figures 2.19, 2.20, 2.21, and 2.22, single-metal sense amps are laid out at half pitch: one amplifier for every two array digitlines. This type of layout is extremely difficult and places tight constraints on process design margins. With the loss of Metal2, the column select signals are not brought across the memory arrays. Generating column select signals locally for each set of I/O transistors requires a full column decode block.

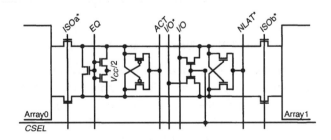

Figure 2.22 Minimum sense amplifier block.

Sec. 2.2 The Sense Amp

As shown in Figure 2.23, the Nsense-amp and Psense-amps are placed on separate ends of the arrays. Sharing sense amps between arrays is especially beneficial for single-metal designs. As illustrated in Figure 2.23, two Psense-amps share a single Nsense-amp. In this case, with the I/O devices on only one end, the right Psense-amp is activated only when the right array is accessed. Conversely, the left Psense-amp is always activated, regardless of the accessed array, because all Read and Write data must pass through the left Psense-amp to get to the I/O devices.

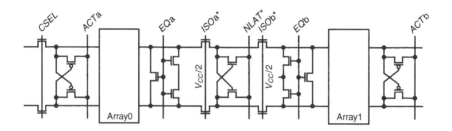

Figure 2.23 Single-metal sense amplifier block.

2.2.7 Operation

A set of signal waveforms is illustrated in Figure 2.24 for the sense amplifier of Figure 2.19. These waveforms depict a Read-Modify-Write cycle (Late Write) in which the cell data is first read out and then new data is written back. In this example, a one level is read out of the cell, as indicated by D^* rising above D during cell access. A one level is always $+V_{CC}/2$ in the mbit cell, regardless of whether it is connected to a true or complement digitline. The correlation between mbit cell data and the data appearing at the DRAM's data terminal *(DQ)* is a function of the data topology and the presence of data scrambling. Data or topo scrambling is implemented at the circuit level: it ensures that the mbit data state and the *DQ* logic level are in agreement. An mbit one level ($+V_{CC}/2$) corresponds to a logic one at the *DQ*, and an mbit zero level ($-V_{CC}/2$) corresponds to a logic zero at the *DQ* terminal.

Writing specific data patterns into the memory arrays is important to DRAM testing. Each type of data pattern identifies the weaknesses or sensitivities of each cell to the data in surrounding cells. These patterns include solids, row stripes, column stripes, diagonals, checkerboards, and a variety of moving patterns. Test equipment must be programmed with the data topology of each type of DRAM to correctly write each pattern. Often the tester itself guarantees that the pattern is correctly written into the arrays, unscrambling the complicated data and address topology as necessary to write a specific pattern. On some newer DRAM designs, part of this task is implemented on the DRAM itself, in the form of a topo scrambler, such that the mbit data state matches the *DQ* logic level. This implementation somewhat simplifies tester programming.

Returning to Figure 2.24, we see that a wordline is fired in Array1. Prior to this, *ISO*a* will go LOW to isolate Array0, and *EQ*b will go LOW to disable the *EQ*/bias transistors connected to Array1. The wordline then fires HIGH, accessing an mbit, which dumps charge onto *D0**. *NLAT**, which is initially at $V_{CC}/2$, drives LOW to begin the Sensing operation and pulls *D0* toward ground. Then, *ACT* fires, moving from ground to V_{CC}, activating the Psense-amplifier and driving *D0** toward V_{CC}. After separation has commenced, *CSEL0* rises to V_{CC}, turning ON the I/O transistors so that the cell data can be read by peripheral circuits. The I/O lines are biased at a voltage close to V_{CC}, which causes *D1* to rise while the column is active. After the Read is complete, Write drivers in the periphery turn ON and drive the I/O lines to opposite data states (in our example).

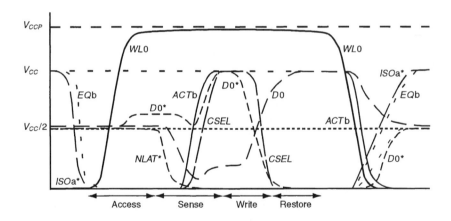

Figure 2.24 Waveforms for the Read-Modify-Write cycle.

This new data propagates through the I/O devices and writes over the existing data stored in the sense amplifiers. Once the sense amplifiers latch the new data, the Write drivers and the I/O devices can be shut down, allowing the sense amplifiers to finish restoring the digitlines to full levels. Following this restoration, the wordline transitions LOW to shut OFF the mbit transistor. Finally, *EQ*b and *ISO*a* fire HIGH to equilibrate the digitlines

Sec. 2.3 Row Decoder Elements 57

back to $V_{CC}/2$ and reconnect Array0 to the sense amplifiers. The timing for each event of Figure 2.24 depends on circuit design, transistor sizes, layout, device performance, parasitics, and temperature. While timing for each event must be minimized to achieve optimum DRAM performance, it cannot be pushed so far as to eliminate all timing margins. Margins are necessary to ensure proper device operation over the expected range of process variations and the wide range of operating conditions.

Again, there is not one set of timing waveforms that covers all design options. The sense amps of Figures 2.19–2.23 all require slightly different signals and timings. Various designs actually fire the Psense-amplifier prior to or coincident with the Nsense-amplifier. This obviously places greater constraints on the Psense-amplifier design and layout, but these constraints are balanced by potential performance benefits. Similarly, the sequence of events as well as the voltages for each signal can vary. There are almost as many designs for sense amplifier blocks as there are DRAM design engineers. Each design reflects various influences, preconceptions, technologies, and levels of understanding. The bottom line is to maximize yield and performance and minimize everything else.

2.3 ROW DECODER ELEMENTS

Row decode circuits are similar to sense amplifier circuits in that they pitch up to mbit arrays and have a variety of implementations. A row decode block is comprised of two basic elements: a wordline driver and an address decoder tree. There are three basic configurations for wordline driver circuits: the NOR driver, the inverter (CMOS) driver, and the bootstrap driver. In addition, the drivers and associated decode trees can be configured either as local row decodes for each array section or as global row decodes that drive a multitude of array sections.

Global row decodes connect to multiple arrays through metal wordline straps. The straps are stitched to the polysilicon wordlines at specific intervals dictated by the polysilicon resistance and the desired *RC* wordline time constant. Most processes that strap wordlines with metal do not silicide the polysilicon, although doing so would reduce the number of stitch regions required. Strapping wordlines and using global row decoders obviously reduces die size [22], very dramatically in some cases. The disadvantage of strapping is that it requires an additional metal layer at minimum array pitch. This puts a tremendous burden on process technologists in which three conductors are at minimum pitch: wordlines, digitlines, and wordline straps.

Local row decoders, on the other hand, require additional die size rather than metal straps. It is highly advantageous to reduce the polysilicon resistance in order to stretch the wordline length and lower the number of row decodes needed. This is commonly achieved with silicided polysilicon processes. On large DRAMs, such as the 1Gb, the area penalty can be prohibitive, making low-resistance wordlines all the more necessary.

2.3.1 Bootstrap Wordline Driver

The bootstrap wordline driver shown in Figure 2.25 is built exclusively from NMOS transistors [22], resulting in the smallest layout of the three types of driver circuits. The absence of PMOS transistors eliminates large Nwell regions from the layout. As the name denotes, this driver relies on bootstrapping principles to bias the output transistor's gate terminal. This bias voltage must be high enough to allow the NMOS transistor to drive the wordline to the boosted wordline voltage V_{CCP}.

Operation of the bootstrap driver is described with help from Figure 2.26. Initially, the driver is OFF with the wordline and *PHASE* terminals at ground. The wordline is held at ground by transistor M2 because the decoder output signal *DEC** is at V_{CC}. The gate of M3, the pass transistor, is fixed at V_{CC}. The signals *DEC* and *DEC** are fed from a decode circuit that will be discussed later. As a complement pair, *DEC* and *DEC** represent the first of two terms necessary to decode the correct wordline. *PHASE0*, which is also fed from a decode circuit, represents the second term. If *DEC* rises to V_{CC} and *DEC** drops to ground, as determined by the decoder, the boot node labeled *B1* will rise to $V_{CC} - V_{TN}$ V, and M2 will turn OFF. The wordline continues to be held to ground by M1 because *PHASE0* is still grounded. After *B1* rises to $V_{CC} - V_{TN}$, the *PHASE0* signal fires to the boosted wordline voltage V_{CCP}. Because of the gate-to-drain and gate-to-source capacitance of M1, the gate of M1 boots to an elevated voltage, V_{boot}. This voltage is determined by the parasitic capacitance of node *B1*, C_{GS1}, C_{GD1}, V_{CCP}, and the initial voltage at *B1*, $V_{CC} - V_{TN}$. Accordingly,

$$V_{boot} \cong \frac{(V_{CCP} \cdot C_{GD1})}{(C_{GS1} + C_{GD1} + C_{B1})} + (V_{CC} - V_{TN}).$$

In conjunction with the wordline voltage rising from ground to V_{CCP}, the gate-to-source capacitance of M1 provides a secondary boost to the boot node. The secondary boost helps to ensure that the boot voltage is adequate to drive the wordline to a full V_{CCP} level.

Sec. 2.3 Row Decoder Elements

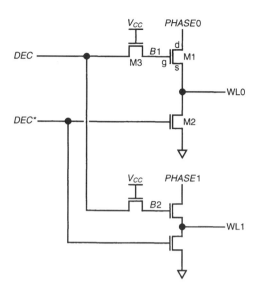

Figure 2.25 Bootstrap wordline driver.

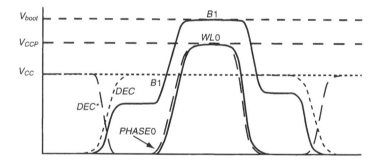

Figure 2.26 Bootstrap operation waveforms.

The layout of the boot node is very important to the bootstrap wordline driver. First, the parasitic capacitance of node $B1$, which includes routing, junction, and overlap components, must be minimized to achieve maximum boot efficiency. Second, charge leakage from the boot node must be minimized to ensure adequate V_{GS} for transistor M1 such that the wordline remains at V_{CCP} for the maximum $\overline{RAS}$ low period. Low leakage is often achieved by minimizing the source area for M3 or using donut gate structures that surround the source area, as illustrated in Figure 2.27.

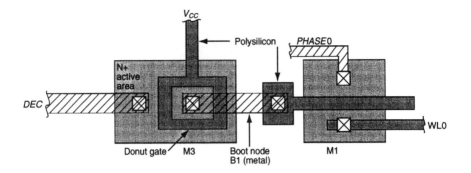

Figure 2.27 Donut gate structure layout.

The bootstrap driver is turned OFF by first driving the *PHASE0* signal to ground. M1 remains ON because node *B1* cannot drop below $V_{CC} - V_{TH}$; M1 substantially discharges the wordline toward ground. This is followed by the address decoder turning OFF, bringing *DEC* to ground and *DEC** to V_{CC}. With *DEC** at V_{CC}, transistor M2 turns ON and fully clamps the wordline to ground. A voltage level translator is required for the *PHASE* signal because it operates between ground and the boosted voltage V_{CCP}. For a global row decode configuration, this requirement is not much of a burden. For a local row decode configuration, however, the requirement for level translators can be very troublesome. Generally, these translators are placed either in the array gap cells at the intersection of the sense amplifier blocks and row decode blocks or distributed throughout the row decode block itself. The translators require both PMOS and NMOS transistors and must be capable of driving large capacitive loads. Layout of the translators is exceedingly difficult, especially because the overall layout needs to be as small as possible.

2.3.2 NOR Driver

The second type of wordline driver is similar to the bootstrap driver in that two decode terms drive the output transistor from separate terminals. The NOR driver, as shown in Figure 2.28, uses a PMOS transistor for M1 and does not rely on bootstrapping to derive the gate bias. Rather, the gate is driven by a voltage translator that converts *DEC* from V_{CC} to V_{CCP} voltage levels. This conversion is necessary to ensure that M1 remains OFF for unselected wordlines as the *PHASE* signal, which is shared by multiple drivers, is driven to V_{CCP}.

Sec. 2.3 Row Decoder Elements   61

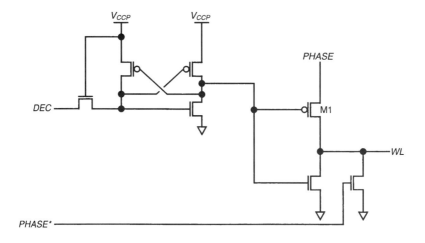

Figure 2.28 NOR driver.

To fire a specific wordline, *DEC* must be HIGH and the appropriate *PHASE* must fire HIGH. Generally, there are four to eight *PHASE* signals per row decoder block. The NOR driver requires one level translator for each *PHASE* and *DEC* signal. By comparison, the bootstrap driver only requires level translators for the *PHASE* signal.

2.3.3 CMOS Driver

The final wordline driver configuration is shown in Figure 2.29. In general, it lacks a specific name: it is sometimes referred to as a CMOS inverter driver or a CMOS driver. Unlike the first two drivers, the output transistor M1 has its source terminal permanently connected to V_{CCP}. This driver, therefore, features a voltage translator for each and every wordline. Both decode terms *DEC* and *PHASE** combine to drive the output stage through the translator. The advantage of this driver, other than simple operation, is low power consumption. Power is conserved because the translators drive only the small capacitance associated with a single driver. The *PHASE* translators of both the bootstrap and NOR drivers must charge considerable junction capacitance. The disadvantages of the CMOS driver are layout complexity and standby leakage current. Standby leakage current is a product of V_{CCP} voltage applied to M1 and its junction and subthreshold leakage currents. For a large DRAM with high numbers of wordline drivers, this leakage current can easily exceed the entire standby current budget unless great care is exercised in designing output transistor M1.

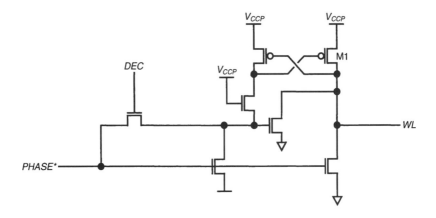

Figure 2.29 CMOS driver.

2.3.4 Address Decode Trees

With the wordline driver circuits behind us, we can turn our attention to the address decoder tree. There is no big secret to address decoding in the row decoder network. Just about any type of logic suffices: static, dynamic, pass gate, or a combination thereof. With any type of logic, however, the primary objectives in decoder design are to maximize speed and minimize die area. Because a great variety of methods have been used to implement row address decoder trees, it is next to impossible to cover them all. Instead, we will give an insight into the possibilities by discussing a few of them.

Regardless of the type of logic with which a row decoder is implemented, the layout must completely reside beneath the row address signal lines to constitute an efficient, minimized design. In other words, the metal address tracks dictate the die area available for the decoder. Any additional tracks necessary to complete the design constitute wasted silicon. For DRAM designs requiring global row decode schemes, the penalty for inefficient design may be insignificant; however, for distributed local row decode schemes, the die area penalty may be significant. As with mbits and sense amplifiers, time spent optimizing row decode circuits is time well spent.

2.3.5 Static Tree

The most obvious form of address decode tree uses static CMOS logic. As shown in Figure 2.30, a simple tree can be designed using two-input NAND gates. While easy to design schematically, static logic address trees are not popular. They waste silicon and are very difficult to lay out efficiently. Static logic requires two transistors for each address term, one

NMOS and one PMOS, which can be significant for many address terms. Furthermore, static gates must be cascaded to accumulate address terms, adding gate delays at each level. For these and other reasons, static logic gates are not used in row decode address trees in today's state-of-the-art DRAMs.

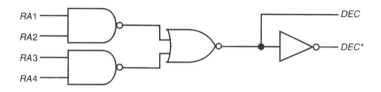

Figure 2.30 Static decode tree.

2.3.6 P&E Tree

The second type of address tree uses dynamic logic, the most prevalent being *precharge and evaluate* (P&E) logic. Used by the majority of DRAM manufacturers, P&E address trees come in a variety of configurations, although the differences between them can be subtle. Figure 2.31 shows a simplified schematic for one version of a P&E address tree designed for use with bootstrapped wordline drivers. P&E address tree circuits feature one or more PMOS *PRECHARGE* transistors and a cascade of NMOS *ENABLE* transistors M2–M4. This P&E design uses half of the transistors required by the static address tree of Figure 2.30. As a result, the layout of the P&E tree is much smaller than that of the static tree and fits more easily under the address lines. The *PRE* transistor is usually driven by a *PRECHARGE** signal under the control of the $\overline{RAS}$ chain logic. *PRECHARGE** and transistor M1 ensure that *DEC** is precharged HIGH, disabling the wordline driver and preparing the tree for row address activation.

M7 is a weak PMOS transistor driven by the *DEC* inverter (M5 and M6). Together, M7 and the inverter form a latch to ensure that *DEC** remains HIGH for all decoders that are not selected by the row addresses. At the beginning of any $\overline{RAS}$ cycle, *PRECHARGE** is LOW and the row addresses are all disabled (LOW). After $\overline{RAS}$ falls, *PRECHARGE** transitions HIGH to turn OFF M1; then the row addresses are enabled. If *RA1–RA3* all go HIGH, then M2–M4 turn ON, overpowering M7 and driving *DEC** to ground and subsequently *DEC* to V_{CC}. The output of this tree segment normally drives four bootstrapped wordline drivers, each connected to a separate *PHASE* signal. Therefore, for an array with 256 wordlines, there will be 64 such decode trees.

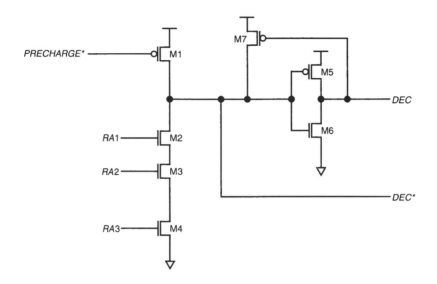

Figure 2.31 P&E decode tree.

2.3.7 Predecoding

The row address lines shown as *RA1–RA3* can be either true and complement or predecoded. Predecoded address lines are formed by logically combining (AND) addresses as shown in Table 2.1.

Table 2.1 Predecoded address truth table.

RA0	RA1	PR01<n>	PR01<0>	PR01<1>	PR01<2>	PR01<3>
0	0	0	1	0	0	0
1	0	1	0	1	0	0
0	1	2	0	0	1	0
1	1	3	0	0	0	1

The advantages of using predecoded addresses include lower power (fewer signals make transitions during address changes) and higher efficiency (only three transistors are necessary to decode six addresses for the circuit of Figure 2.31). Predecoding is especially beneficial in redundancy circuits. In fact, predecoded addresses are used throughout most DRAM designs today.

2.3.8 Pass Transistor Tree

The final type of address tree to be examined, shown in Figure 2.32, uses pass transistor logic. Pass transistor address trees are similar to P&E trees in numerous ways. Both designs use PMOS *PRECHARGE* transistors and NMOS address *ENABLE* transistors. Unlike P&E logic, however, the NMOS cascade does not terminate at ground. Rather, the cascade of M2–M4 goes to a *PHASE** signal, which is HIGH during *PRECHARGE* and LOW during *EVALUATE*. The address signals operate the same as in the P&E tree: HIGH to select and LOW to deselect. The pass transistor tree is shown integrated into a CMOS wordline driver. This is necessary because the pass transistor tree and the CMOS wordline driver are generally used together and their operation is complementary. The cross-coupled PMOS transistors of the CMOS level translator provide a latch necessary to keep the final interstage node biased at V_{CC}. Again the latch has a weak pull-up, which is easily overpowered by the cascaded NMOS *ENABLE* transistors. A pass transistor address tree is not used with bootstrapped wordline drivers because the *PHASE* signal feeds into the address tree logic rather than into the driver, as required by the bootstrap driver.

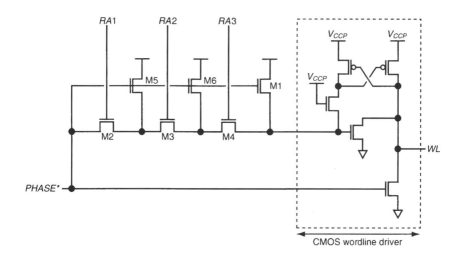

Figure 2.32 Pass transistor decode tree.

2.4 DISCUSSION

We have briefly examined the basic elements required in DRAM row decoder blocks. Numerous variations are possible. No single design is best for all applications. As with sense amplifiers, design depends on technology and performance and cost trade-offs.

REFERENCES

[1] K. Itoh, "Trends in Megabit DRAM Circuit Design," *IEEE Journal of Solid-State Circuits,* vol. 25, pp. 778–791, June 1990.

[2] D. Takashima, S. Watanabe, H. Nakano, Y. Oowaki, and K. Ohuchi, "Open/Folded Bit-Line Arrangement for Ultra-High-Density DRAM's," *IEEE Journal of Solid-State Circuits,* vol. 29, pp. 539–542, April 1994.

[3] Hideto Hidaka, Yoshio Matsuda, and Kazuyasu Fujishima, "A Divided/Shared Bit-Line Sensing Scheme for ULSI DRAM Cores," *IEEE Journal of Solid-State Circuits,* vol. 26, pp. 473–477, April 1991.

[4] M. Aoki, Y. Nakagome, M. Horiguchi, H. Tanaka, S. Ikenaga, J. Etoh, Y. Kawamoto, S. Kimura, E. Takeda, H. Sunami, and K. Itoh, "A 60-ns 16-Mbit CMOS DRAM with a Transposed Data-Line Structure," *IEEE Journal of Solid-State Circuits,* vol. 23, pp. 1113–1119, October 1988.

[5] R. Kraus and K. Hoffmann, "Optimized Sensing Scheme of DRAMs," *IEEE Journal of Solid-State Circuits,* vol. 24, pp. 895–899, August 1989.

[6] T. Yoshihara, H. Hidaka, Y. Matsuda, and K. Fujishima, "A Twisted Bitline Technique for Multi-Mb DRAMs," *1988 IEEE ISSCC Digest of Technical Papers,* pp. 238–239.

[7] Yukihito Oowaki, Kenji Tsuchida, Yohji Watanabe, Daisaburo Takashima, Masako Ohta, Hiroaki Nakano, Shigeyoshi Watanabe, Akihiro Nitayama, Fumio Horiguchi, Kazunori Ohuchi, and Fujio Masuoka, "A 33-ns 64Mb DRAM," *IEEE Journal of Solid-State Circuits,* vol. 26, pp. 1498–1505, November 1991.

[8] M. Inoue, H. Kotani, T. Yamada, H. Yamauchi, A. Fujiwara, J. Matsushima, H. Akamatsu, M. Fukumoto, M. Kubota, I. Nakao, N. Aoi, G. Fuse, S. Ogawa, S. Odanaka. A. Ueno, and Y. Yamamoto, "A 16Mb DRAM with an Open Bit-Line Architecture," *1988 IEEE ISSCC Digest of Technical Papers,* pp. 246–247.

[9] Y. Kubota, Y. Iwase, K. Iguchi, J. Takagi, T. Watanabe, and K. Sakiyama, "Alternatively-Activated Open Bitline Technique for High Density DRAM's," *IEICE Trans. Electron.,* vol. E75-C, pp. 1259–1266, October 1992.

[10] T. Hamada, N. Tanabe, H. Watanabe, K. Takeuchi, N. Kasai, H. Hada, K. Shibahara, K. Tokashiki, K. Nakajima, S. Hirasawa, E. Ikawa, T. Saeki, E. Kakehashi, S. Ohya, and T. Kunio, "A Split-Level Diagonal Bit-Line (SLDB) Stacked Capacitor Cell for 256Mb DRAMs," *1992 IEDM Technical Digest,* pp. 799–802.

[11] Toshinori Morihara, Yoshikazu Ohno, Takahisa Eimori, Toshiharu Katayama, Shinichi Satoh, Tadashi Nishimura, and Hirokazu Miyoshi, "Disk-Shaped Stacked Capacitor Cell for 256Mb Dynamic Random-Access Memory,"

Japan Journal of Applied Physics, vol. 33, Part 1, pp. 4570–4575, August 1994.

[12] J. H. Ahn, Y. W. Park, J. H. Shin, S. T. Kim, S. P. Shim, S. W. Nam, W. M. Park, H. B. Shin, C. S. Choi, K. T. Kim, D. Chin, O. H. Kwon, and C. G. Hwang, "Micro Villus Patterning (MVP) Technology for 256Mb DRAM Stack Cell," 1992 Symposium on VLSI Tech. Digest of Technical Papers, pp. 12–13.

[13] Kazuhiko Sagara, Tokuo Kure, Shoji Shukuri, Jiro Yugami, Norio Hasegawa, Hidekazu Goto, and Hisaomi Yamashita, "Recessed Memory Array Technology for a Double Cylindrical Stacked Capacitor Cell of 256M DRAM," *IEICE Trans. Electron.,* vol. E75-C, pp. 1313–1322, November 1992.

[14] S. Ohya, "Semiconductor Memory Device Having Stacked-Type Capacitor of Large Capacitance," United States Patent Number 5,298,775, March 29, 1994.

[15] M. Sakao, N. Kasai, T. Ishijima, E. Ikawa, H. Watanabe, K. Terada, and T. Kikkawa, "A Capacitor-Over-Bit-Line (COB) Cell with Hemispherical-Grain Storage Node for 64Mb DRAMs," *1990 IEDM Technical Digest,* pp. 655–658.

[16] G. Bronner, H. Aochi, M. Gall, J. Gambino, S. Gernhardt, E. Hammerl, H. Ho, J. Iba, H. Ishiuchi, M. Jaso, R. Kleinhenz, T. Mii, M. Narita, L. Nesbit, W. Neumueller, A. Nitayama, T. Ohiwa, S. Parke, J. Ryan, T. Sato, H. Takato, and S. Yoshikawa, "A Fully Planarized 0.25µm CMOS Technology for 256Mbit DRAM and Beyond," 1995 Symposium on VLSI Tech. Digest of Technical Papers, pp. 15–16.

[17] N. C.-C. Lu and H. H. Chao, "Half-V_{DD} / Bit-Line Sensing Scheme in CMOS DRAMs," in *IEEE Journal of Solid-State Circuits,* vol. SC19, p. 451, August 1984.

[18] E. Adler; J. K. DeBrosse; S. F. Geissler; S. J. Holmes; M. D. Jaffe; J. B. Johnson; C. W. Koburger, III; J. B. Lasky; B. Lloyd; G. L. Miles; J. S. Nakos; W. P. Noble, Jr.; S. H. Voldman; M. Armacost; and R. Ferguson; "The Evolution of IBM CMOS DRAM Technology;" *IBM Journal of Research and Development,* vol. 39, pp. 167–188, March 1995.

[19] R. Kraus, "Analysis and Reduction of Sense-Amplifier Offset," in *IEEE Journal of Solid-State Circuits,* vol. 24, pp. 1028–1033, August 1989.

[20] M. Asakura, T. Ohishi, M. Tsukude, S. Tomishima, H. Hidaka, K. Arimoto, K. Fujishima, T. Eimori, Y. Ohno, T. Nishimura, M. Yasunaga, T. Kondoh, S. I. Satoh, T. Yoshihara, and K. Demizu, "A 34ns 256Mb DRAM with Boosted Sense-Ground Scheme," *1994 IEEE ISSCC Digest of Technical Papers,* pp. 140–141.

[21] T. Ooishi, K. Hamade, M. Asakura, K. Yasuda, H. Hidaka, H. Miyamoto, and H. Ozaki, "An Automatic Temperature Compensation of Internal Sense Ground for Sub-Quarter Micron DRAMs," 1994 Symposium on VLSI Circuits Digest of Technical Papers, pp. 77–78.

[22] K. Noda, T. Saeki, A. Tsujimoto, T. Murotani, and K. Koyama, "A Boosted Dual Word-line Decoding Scheme for 256 Mb DRAMs," 1992 Symposium on VLSI Circuits Digest of Technical Papers, pp. 112–113.

Chapter 3

Array Architectures

This chapter presents a detailed description of the two most prevalent array architectures under consideration for future large-scale DRAMs: the aforementioned open architectures and folded digitline architectures.

3.1 ARRAY ARCHITECTURES

To provide a viable point for comparison, each architecture is employed in the theoretical construction of 32-Mbit memory blocks for use in a 256-Mbit DRAM. Design parameters and layout rules from a typical 0.25 µm DRAM process provide the necessary dimensions and constraints for the analysis. Some of these parameters are shown in Table 3.1. By examining DRAM architectures in the light of a real-world design problem, an objective and unbiased comparison can be made. In addition, using this approach, we readily detect the strengths and weaknesses of either architecture.

3.1.1 Open Digitline Array Architecture

The open digitline array architecture was the prevalent architecture prior to the 64kbit DRAM. A modern embodiment of this architecture, as shown in Figure 3.1 [1] [2], is constructed with multiple crosspoint array cores separated by strips of sense amplifier blocks in one axis and either row decode blocks or wordline stitching regions in the other axis. Each 128kbit array core is built using $6F^2$ mbit cell pairs. There are 131,072 (2^{17}) functionally addressable mbits arranged in 264 rows and 524 digitlines. In the 264 rows, there are 256 actual wordlines, 4 redundant wordlines, and 4 dummy wordlines. In the 524 digitlines, there are 512 actual digitlines, 8 redundant digitlines, and 4 dummy digitlines. Photolithography problems usually occur at the edge of large repetitive structures, such as mbit arrays. These problems produce malformed or nonuniform structures, rendering the edge cells useless. Therefore, including dummy mbits, wordlines, and digitlines on each array edge ensures that these problems occur only on dummy cells, leaving live cells unaffected. Although dummy structures enlarge each array core, they also significantly improve device yield. Thus, they are necessary on all DRAM designs.

Table 3.1 0.25 µm design parameters.

Parameter	Value
Digitline width W_{DL}	0.3 µm
Digitline pitch P_{DL}	0.6 µm
Wordline width W_{WL}	0.3 µm
Wordline pitch for $8F^2$ mbit P_{WL8}	0.6 µm
Wordline pitch for $6F^2$ mbit P_{WL6}	0.9 µm
Cell capacitance C_C	30fF
Digitline capacitance per mbit C_{DM}	0.8fF
Wordline capacitance per $8F^2$ mbit C_{W8}	0.6fF
Wordline capacitance per $6F^2$ mbit C_{W6}	0.5fF
Wordline sheet resistance R_S	6Ω/sq

Array core size, as measured in the number of mbits, is restricted by two factors: a desire to keep the quantity of mbits a power of two and the practical limits on wordline and digitline length. The need for a binary quantity of mbits in each array core derives from the binary nature of DRAM addressing. Given N row addresses and M column addresses for a given part, there are 2^{N+M} addressable mbits. Address decoding is greatly simplified within a DRAM if array address boundaries are derived directly from address bits.

Sec. 3.1 Array Architectures 71

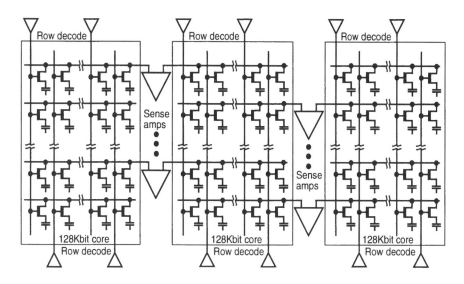

Figure 3.1 Open digitline architecture schematic.

Because addressing is binary, the boundaries naturally become binary. Therefore, the size of each array core must necessarily have 2^X addressable rows and 2^Y addressable digitlines. The resulting array core size is 2^{X+Y} mbits, which is, of course, a binary number.

The second set of factors limiting array core size involves practical limits on digitline and wordline length. From earlier discussions in Section 2.1, the digitline capacitance is limited by two factors. First, the ratio of cell capacitance to digitline capacitance must fall within a specified range to ensure reliable sensing. Second, operating current and power for the DRAM is in large part determined by the current required to charge and discharge the digitlines during each active cycle. Power considerations restrict digitline length for the 256-Mbit generation to approximately 128 mbit pairs (256 rows), with each mbit connection adding capacitance to the digitline. The power dissipated during a Read or Refresh operation is proportional to the *digitline capacitance* (C_D), the supply voltage (V_{CC}), the number of active columns (N), and the Refresh period (P). Accordingly, the power dissipated is given as

$$P_D = V_{CC}2 \cdot \frac{N(C_D + C_C)}{2P} \; watts.$$

On a 256-Mbit DRAM in 8k (rows) Refresh, there are 32,768 (2^{15}) active columns during each Read, Write, or Refresh operation. The active array current and power dissipation for a 256-Mbit DRAM appear in Table 3.2 for a 90 ns Refresh period (–5 timing) at various digitline lengths. The budget for the active array current is limited to 200mA for this 256-Mbit design. To meet this budget, the digitline cannot exceed a length of 256 mbits.

Wordline length, as described in Section 2.1, is limited by the maximum allowable RC time constant of the wordline. To ensure acceptable access time for the 256-Mbit DRAM, the wordline time constant should be kept below 4 nanoseconds. For a wordline connected to N mbits, the total resistance and capacitance follow:

$$R_{WL} = \frac{(R_S \cdot N \cdot P_{DL})}{W_{LW}} \text{ and}$$

$$C_{WL} = C_{W8} \cdot N \text{ farads},$$

where P_{DL} is the digitline pitch, W_{LW} is the wordline width, and C_{W8} is the wordline capacitance in an 8F² mbit cell.

Table 3.2 Active current and power versus digitline length.

Digitline Length (mbits)	Digitline Capacitance (fF)	Active Current (mA)	Power Dissipation (mW)
128	102	60	199
256	205	121	398
512	410	241	795

Table 3.3 contains the effective wordline time constants for various wordline lengths. As shown in the table, the wordline length cannot exceed 512 mbits (512 digitlines) if the wordline time constant is to remain under 4 nanoseconds.

The open digitline architecture does not support digitline twisting because the true and complement digitlines, which constitute a column, are in separate array cores. Therefore, no silicon area is consumed for twist regions. The 32-Mbit array block requires a total of two hundred fifty-six 128kbit array cores in its construction. Each 32-Mbit block represents an address space comprising a total of 4,096 rows and 8,192 columns. A practical configuration for the 32-Mbit block is depicted in Figure 3.2.

Sec. 3.1 Array Architectures

Table 3.3 Wordline time constant versus wordline length.

Wordline Length (mbits)	R_{WL} (ohms)	C_{WL} (fF)	Time Constant (ns)
128	1,536	64	0.098
256	3,072	128	0.39
512	6,144	256	1.57
1024	12,288	512	6.29

In Figure 3.2, the 256 array cores appear in a 16 x 16 arrangement. The x16 arrangement produces 2-Mbit sections of 256 wordlines and 8,192 digit-lines (4,096 columns). Sixteen 2-Mbit sections are required to form the complete 32-Mbit block: sense amplifier strips are positioned vertically between each 2-Mbit section, and row decode strips or wordline stitching strips are positioned horizontally between each array core.

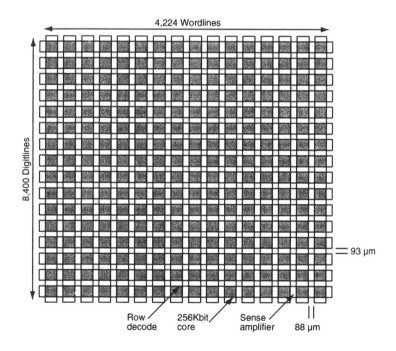

Figure 3.2 Open digitline 32-Mbit array block.

Layout can be generated for the various 32-Mbit elements depicted in Figure 3.2. This layout is necessary to obtain reasonable estimates for pitch cell size. With these size estimates, overall dimensions of the 32-Mbit memory block can be calculated. The results of these estimates appear in Table 3.4. Essentially, the overall height of the 32-Mbit block can be found by summing the height of the row decode blocks (or stitch regions) together with the product of the digitline pitch and the total number of digitlines.

Accordingly,

$$Height32 = (T_R \cdot H_{LDEC}) + (T_{DL} \cdot P_{DL}) \; microns,$$

where T_R is the number of local row decoders, H_{LDEC} is the height of each decoder, T_{DL} is the number of digitlines including redundant and dummy lines, and P_{DL} is the digitline pitch. Similarly, the width of the 32-Mbit block is found by summing the total width of the sense amplifier blocks with the product of the wordline pitch and the number of wordlines. This bit of math yields

$$Width32 = (T_{SA} \cdot W_{AMP}) + (T_{WL} \cdot P_{WL6}) \; microns,$$

where T_{SA} is the number of sense amplifier strips, W_{AMP} is the width of the sense amplifiers, T_{WL} is the total number of wordlines including redundant and dummy lines, and P_{WL6} is the wordline pitch for the 6F^2 mbit.

Table 3.4 contains calculation results for the 32-Mbit block shown in Figure 3.2. Although overall size is the best measure of architectural efficiency, a second popular metric is array efficiency. Array efficiency is determined by dividing the area consumed by functionally addressable mbits by the total die area. To simplify the analysis in this book, peripheral circuits are ignored in the array efficiency calculation. Rather, the calculation considers only the 32-Mbit memory block, ignoring all other factors. With this simplification, the array efficiency for a 32-Mbit block is given as

$$Efficiency = \frac{(100 \cdot 2^{25} \cdot P_{DL} \cdot P_{WL6})}{Area32} \; percent,$$

where 2^{25} is the number of addressable mbits in each 32-Mbit block. The open digitline architecture yields a calculated array efficiency of 51.7%.

Unfortunately, the ideal open digitline architecture presented in Figure 3.2 is difficult to realize in practice. The difficulty stems from an interdependency between the memory array and sense amplifier layouts in which each array digitline must connect to one sense amplifier and each sense amplifier must connect to two array digitlines.

Sec. 3.1 Array Architectures 75

Table 3.4 Open digitline (local row decode)—32-Mbit size calculations.

Description	Parameter	Size
Number of sense amplifier strips	T_{SA}	17
Width of sense amplifiers	W_{AMP}	88 µm
Number of local decode strips	T_{LDEC}	17
Height of local decode strips	H_{LDEC}	93 µm
Number of digitlines	T_{DL}	8,400
Number of wordlines	T_{WL}	4,224
Height of 32-Mbit block	$Height32$	6,621 µm
Width of 32-Mbit block	$Width32$	5,298 µm
Area of 32-Mbit block	$Area32$	35,078,058 µm²

This interdependency, which exists for all array architectures, becomes problematic for the open digitline architecture. The two digitlines, which connect to a sense amplifier, must come from two separate memory arrays. As a result, sense amplifier blocks must always be placed between memory arrays for open digitline array architectures [3], unlike the depiction in Figure 3.2.

Two layout approaches may be used to achieve this goal. First, design the sense amplifiers so that the sense amplifier block contains a set of sense amplifiers for each digitline in the array. This single-pitch solution, shown in Figure 3.3, eliminates the need for sense amplifiers on both sides of an array core because all of the digitlines connect to a single sense amplifier block. Not only does this solution eliminate the edge problem, but it also reduces the 32-Mbit block size. There are now only eight sense amplifier strips instead of the seventeen of Figure 3.2. Unfortunately, it is nearly impossible to lay out sense amplifiers in this fashion [4]. Even a single-metal sense amplifier layout, considered the tightest layout in the industry, achieves only one sense amplifier for every two digitlines (double-pitch).

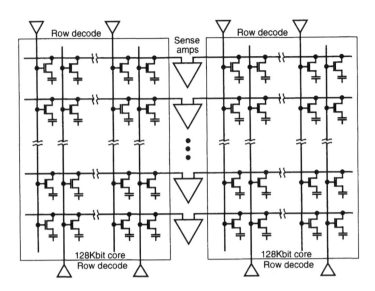

Figure 3.3 Single-pitch open digitline architecture.

A second approach to solving the interdependency problem in open digitline architectures is to maintain the configuration shown in Figure 3.2 but include some form of reference digitline for the edge sense amplifiers. The reference digitline can assume any form as long as it accurately models the capacitance and behavior of a true digitline. Obviously, the best type of reference digitline is a true digitline. Therefore, with this approach, more dummy array cores are added to both edges of the 32-Mbit memory block, as shown in Figure 3.4. The dummy array cores need only half as many wordlines as the true array core because only half of the digitlines are connected to any single sense amplifier strip. The unconnected digitlines double the effective length of the reference digitlines.

This approach solves the array edge problem. However, by producing a larger 32-Mbit memory block, array efficiency is reduced. Dummy arrays solve the array edge problem inherent in open digitline architecture but require sense amplifier layouts that are on the edge of impossible. The problem of sense amplifier layout is all the more difficult because global column select lines must be routed through. For all intents and purposes, therefore, the sense amplifier layout cannot be completed without the presence of an additional conductor, such as a third metal, or without time multiplexed sensing. Thus, for the open digitline architecture to be successful, an additional metal must be added to the DRAM process.

Sec. 3.1 Array Architectures 77

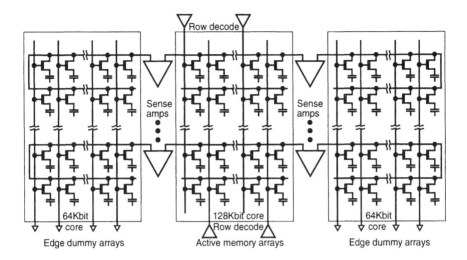

Figure 3.4 Open digitline architecture with dummy arrays.

With the presence of Metal3, the sense amplifier layout and either a full or hierarchical global row decoding scheme is made possible. A full global row decoding scheme using wordline stitching places great demands on metal and contact/via technologies; however, it represents the most efficient use of the additional metal. Hierarchical row decoding using bootstrap wordline drivers is slightly less efficient. Wordlines no longer need to be strapped with metal on pitch, and, thus, process requirements are relaxed significantly [5].

For a balanced perspective, both global and hierarchical approaches are analyzed. The results of this analysis for the open digitline architecture are summarized in Tables 3.5 and 3.6. Array efficiency for global and hierarchical row decoding calculate to 60.5% and 55.9%, respectively, for the 32-Mbit memory blocks based on data from these tables.

Table 3.5 Open digitline (dummy arrays and global row decode)—32-Mbit size calculations.

Description	Parameter	Size
Number of sense amplifier strips	T_{SA}	17
Width of sense amplifiers	W_{AMP}	88 µm
Number of global decode strips	T_{GDEC}	1
Height of global decode strips	H_{GDEC}	200 µm
Number of stitch regions	N_{ST}	17
Height of stitch regions	H_{ST}	10 µm
Number of digitlines	T_{DL}	8,400
Number of wordlines	T_{WL}	4,488
Height of 32-Mbit block	$Height32$	5,410 µm
Width of 32-Mbit block	$Width32$	5,535 µm
Area of 32-Mbit block	$Area32$	29,944,350 µm²

Sec. 3.1 Array Architectures   79

Table 3.6 Open digitline (dummy arrays and hierarchical row decode)—32-Mbit size calculations.

Description	Parameter	Size
Number of sense amplifier strips	T_{SA}	17
Width of sense amplifiers	W_{AMP}	88 µm
Number of global decode strips	T_{GDEC}	1
Height of global decode strips	H_{GDEC}	190 µm
Number of hier decode strips	T_{HDEC}	17
Height of hier decode strips	H_{HDEC}	37 µm
Number of digitlines	T_{DL}	8,400
Number of wordlines	T_{WL}	4,488
Height of 32-Mbit block	*Height*32	5,859 µm
Width of 32-Mbit block	*Width*32	5,535 µm
Area of 32-Mbit block	*Area*32	32,429,565 µm²

3.1.2 Folded Array Architecture

The folded array architecture depicted in Figure 3.5 is the standard architecture of today's modern DRAM designs. The folded architecture is constructed with multiple array cores separated by strips of sense amplifiers and either row decode blocks or wordline stitching regions. Unlike the open digitline architecture, which uses 6F² mbit cell pairs, the folded array core uses 8F² mbit cell pairs [6]. Modern array cores include 262,144 (2^{18}) functionally addressable mbits arranged in 532 rows and 1,044 digitlines. In the 532 rows, there are 512 actual wordlines, 4 redundant wordlines, and 16 dummy wordlines. Each row (wordline) connects to mbit transistors on alternating digitlines. In the 1,044 digitlines, there are 1,024 actual digitlines (512 columns), 16 redundant digitlines (8 columns), and 4 dummy digitlines.

As discussed in Section 3.1.1, because of the additive or subtractive photolithography effects, dummy wordlines and digitlines are necessary to guardband live digitlines. These photo effects are pronounced at the edges of large repetitive structures such as the array cores.

Sense amplifier blocks are placed on both sides of each array core. The sense amplifiers within each block are laid out at quarter pitch: one sense amplifier for every four digitlines. Each sense amplifier connects through isolation devices to columns (digitline pairs) from both adjacent array cores. Odd columns connect on one side of the core, and even columns connect on the opposite side. Each sense amplifier block is therefore connected to only odd or even columns and is never connected to both odd *and* even columns within the same block. Connecting to both odd and even columns requires a half-pitch sense amplifier layout: one sense amplifier for every two digitlines. While half-pitch layout is possible with certain DRAM processes, the bulk of production DRAM designs remain quarter pitch due to the ease of laying them out. The analysis presented in this chapter is accordingly based on quarter-pitch design practices.

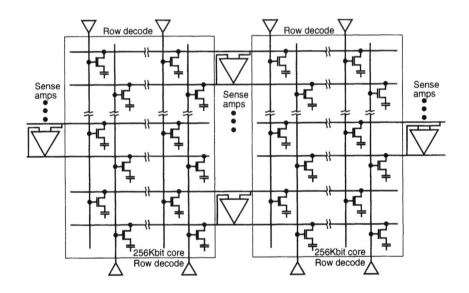

Figure 3.5 Folded digitline array architecture schematic.

The location of row decode blocks for the array core depends on the number of available metal layers. For one- and two-metal processes, local row decode blocks are located at the top and bottom edges of the core. For

Sec. 3.1 Array Architectures

three- and four-metal processes, global row decodes are used. Global row decodes require only stitch regions or local wordline drivers at the top and bottom edges of the core [7]. Stitch regions consume much less silicon area than local row decodes, substantially increasing array efficiency for the DRAM. The array core also includes digitline twist regions running parallel to the wordlines. These regions provide the die area required for digitline twisting. Depending on the particular twisting scheme selected for a design (see Section 2.1), the array core needs between one and three twist regions. For the sake of analysis, a triple twist is assumed, as it offers the best overall noise performance and has been chosen by DRAM manufacturers for advanced large-scale applications [8]. Because each twist region constitutes a break in the array structure, it is necessary to use dummy wordlines. For this reason, there are 16 dummy wordlines (2 for each array edge) in the folded array core rather than 4 dummy wordlines as in the open digitline architecture.

There are more mbits in the array core for folded digitline architectures than there are for open digitline architectures. Larger core size is an inherent feature of folded architectures, arising from the very nature of the architecture. The term *folded architecture* comes from the fact that folding two open digitline array cores one on top of the other produces a folded array core. The digitlines and wordlines from each folded core are spread apart (double pitch) to allow room for the other folded core. After folding, each constituent core remains intact and independent, except for the mbit changes ($8F^2$ conversion) necessary in the folded architecture. The array core size doubles because the total number of digitlines and wordlines doubles in the folding process. It does not quadruple as one might suspect because the two constituent folded cores remain independent: the wordlines from one folded core do not connect to mbits in the other folded core.

Digitline pairing (column formation) is a natural outgrowth of the folding process; each wordline only connects to mbits on alternating digitlines. The existence of digitline pairs (columns) is the one characteristic of folded digitline architectures that produces superior signal-to-noise performance. Furthermore, the digitlines that form a column are physically adjacent to one another. This feature permits various digitline twisting schemes to be used, as discussed in Section 2.1, further improving signal-to-noise performance.

Similar to the open digitline architecture, digitline length for the folded digitline architecture is again limited by power dissipation and minimum cell-to-digitline capacitance ratio. For the 256-Mbit generation, digitlines are restricted from connecting to more than 256 cells (128 mbit pairs). The analysis used to arrive at this quantity is similar to that for the open digitline architecture. (Refer to Table 3.2 to view the calculated results of power dissipation versus digitline length for a 256-Mbit DRAM in 8k Refresh.) Wordline length is again limited by the maximum allowable RC time constant of the wordline.

Contrary to an open digitline architecture in which each wordline connects to mbits on each digitline, the wordlines in a folded digitline architecture connect to mbits only on alternating digitlines. Therefore, a wordline can cross 1,024 digitlines while connecting to only 512 mbit transistors. The wordlines have twice the overall resistance, but only slightly more capacitance because they run over field oxide on alternating digitlines. Table 3.7 presents the effective wordline time constants for various wordline lengths for a folded array core. For a wordline connected to N mbits, the total resistance and capacitance follow:

$$R_{WL} = (2 \cdot R_S \cdot N \cdot P_{DL}) \; ohms$$

$$C_{WL} = C_{W8} \cdot N \; farads,$$

where P_{DL} is the digitline pitch and C_{W8} is the wordline capacitance in an 8F² mbit cell. As shown in Table 3.7, the wordline length cannot exceed 512 mbits (1,024 digitlines) for the wordline time constant to remain under 4 nanoseconds. Although the wordline connects to only 512 mbits, it is two times longer (1,024 digitlines) than wordlines in open digitline array cores. The folded digitline architecture therefore requires half as many row decode blocks or wordline stitching regions as the open digitline architecture.

Table 3.7 Wordline time constant versus wordline length (folded).

Wordline Length (mbits)	R_{WL} (ohms)	C_{WL} (fF)	Time Constant (ns)
128	3,072	77	0.24
256	6,144	154	0.95
512	12,288	307	3.77
1,024	24,576	614	15.09

A diagram of a 32-Mbit array block using folded digitline architecture is shown in Figure 3.6. This block requires a total of one hundred twenty-eight 256kbit array cores. In this figure, the array cores are arranged in an 8-row and 16-column configuration. The x8 row arrangement produces 2-Mbit sec-

Sec. 3.1 Array Architectures

tions of 256 wordlines and 8,192 digitlines (4,096 columns). A total of sixteen 2-Mbit sections form the complete 32-Mbit array block. Sense amplifier strips are positioned vertically between each 2-Mbit section, as in the open digitline architecture. Again, row decode blocks or wordline stitching regions are positioned horizontally between the array cores.

The 32-Mbit array block shown in Figure 3.6 includes size estimates for the various pitch cells. The layout was generated where necessary to arrive at the size estimates. The overall size for the folded digitline 32-Mbit block can be found by again summing the dimensions for each component. Accordingly,

$$Height32 = (T_R \cdot H_{RDEC}) + (T_{DL} \cdot P_{DL}) \; microns,$$

where T_R is the number of row decoders, H_{RDEC} is the height of each decoder, T_{DL} is the number of digitlines including redundant and dummy, and P_{DL} is the digitline pitch. Similarly,

$$Width32 = (T_{SA} \cdot W_{AMP}) + (T_{WL} \cdot P_{WL8}) + (T_{TWIST} \cdot W_{TWIST}) \; microns,$$

where T_{SA} is the number of sense amplifier strips, W_{AMP} is the width of the sense amplifiers, T_{WL} is the total number of wordlines including redundant and dummy, P_{WL8} is the wordline pitch for the 8F^2 mbit, T_{TWIST} is the total number of twist regions, and W_{TWIST} is the width of the twist regions.

Table 3.8 shows the calculated results for the 32-Mbit block shown in Figure 3.6. In this table, a double-metal process is used, which requires local row decoder blocks. Note that Table 3.8 for the folded digitline architecture contains approximately twice as many wordlines as does Table 3.5 for the open digitline architecture. The reason for this is that each wordline in the folded array only connects to mbits on alternating digitlines, whereas each wordline in the open array connects to mbits on every digitline. A folded digitline design therefore needs twice as many wordlines as a comparable open digitline design.

Array efficiency for the 32-Mbit memory block from Figure 3.6 is again found by dividing the area consumed by functionally addressable mbits by the total die area. For the simplified analysis presented in this book, the peripheral circuits are ignored. Array efficiency for the 32-Mbit block is therefore given as

$$Efficiency = \frac{(100 \cdot 2^{25} \cdot P_{DL} \cdot 2 \cdot P_{WL8})}{Area32} \; percent,$$

which yields 59.5% for the folded array design example.

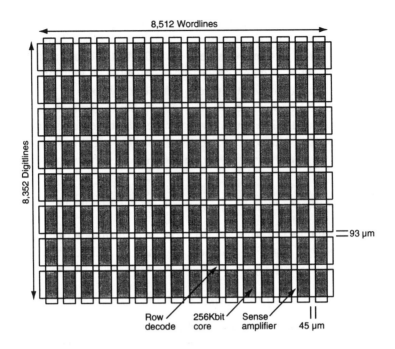

Figure 3.6 Folded digitline architecture 32-Mbit array block.

With the addition of Metal3 to the DRAM process, either a global or hierarchical row decoding scheme, similar to the open digitline analysis, can be used. While global row decoding and stitched wordlines achieve the smallest die size, they also place greater demands on the fabrication process. For a balanced perspective, both approaches were analyzed for the folded digitline architecture. The results of this analysis are presented in Tables 3.9 and 3.10. Array efficiency for the 32-Mbit memory blocks using global and hierarchical row decoding calculate to 74.0% and 70.9%, respectively.

Sec. 3.1 Array Architectures

Table 3.8 Folded digitline (local row decode)—32-Mbit size calculations.

Description	Parameter	Size
Number of sense amplifier strips	T_{SAF}	17
Width of sense amplifiers	W_{AMP}	45 µm
Number of local decode strips	T_{LDEC}	9
Height of local decode strips	H_{LDEC}	93 µm
Number of digitlines	T_{DL}	8,352
Number of wordlines	T_{WL1}	8,512
Number of twist regions	T_{TWIST}	48
Width of twist regions	W_{TWIST}	6 µm
Height of 32-Mbit block	$Height32$	6,592 µm
Width of 32-Mbit block	$Width32$	6,160 µm
Area of 32-Mbit block	$Area32$	40,606,720 µm²

Table 3.9 Folded digitline (global decode)—32-Mbit size calculations.

Description	Parameter	Size
Number of sense amplifier strips	T_{SA}	17
Width of sense amplifiers	W_{AMP}	45 μm
Number of global decode strips	T_{GDEC}	1
Height of global decode strips	H_{GDEC}	200 μm
Number of stitch regions	N_{ST}	9
Height of stitch regions	H_{ST}	10 μm
Number of digitlines	T_{DL}	8,352
Number of wordlines	T_{WL}	8,512
Number of twist regions	T_{TWIST}	48
Width of twist regions	W_{TWIST}	6 μm
Height of 32-Mbit block	$Height32$	5,301 μm
Width of 32-Mbit block	$Width32$	6,160 μm
Area of 32-Mbit block	$Area32$	32,654,160 μm^2

Sec. 3.2 Design Examples: Advanced Bilevel DRAM Architecture   87

Table 3.10 Folded digitline (hierarchical row decode)—32-Mbit size calculations.

Description	Parameter	Size
Number of sense amplifier strips	T_{SA}	17
Width of sense amplifiers	W_{AMP}	45 µm
Number of global decode strips	T_{GDEC}	1
Height of global decode strips	H_{GDEC}	190 µm
Number of hier decode strips	N_{HDEC}	9
Height of hier decode strips	H_{HEC}	37 µm
Number of digitlines	T_{DL}	8,352
Number of wordlines	T_{WL}	8,512
Number of twist regions	T_{TWIST}	48
Width of twist regions	W_{TWIST}	6 µm
Height of 32-Mbit block	$Height32$	5,534 µm
Width of 32-Mbit block	$Width32$	6,160 µm
Area of 32-Mbit block	$Area32$	34,089,440 µm²

3.2 DESIGN EXAMPLES: ADVANCED BILEVEL DRAM ARCHITECTURE

This section introduces a novel, advanced architecture possible for use on future large-scale DRAMs. First, we discuss technical objectives for the proposed architecture. Second, we develop and describe the concept for an advanced array architecture capable of meeting these objectives. Third, we conceptually construct a 32-Mbit memory block with this new architecture for a 256-Mbit DRAM. Finally, we compare the results achieved with the new architecture to those obtained for the open digitline and folded digitline architectures from Section 3.1.

3.2.1 Array Architecture Objectives

Both the open digitline and folded digitline architectures have distinct advantages and disadvantages. While open digitline architectures achieve smaller array layouts by virtue of using smaller $6F^2$ mbit cells, they suffer from poor signal-to-noise performance. A relaxed wordline pitch, which stems from the $6F^2$ mbit, simplifies the task of wordline driver layout. Sense amplifier layout, however, is difficult because the array configuration is inherently half pitch: one sense amplifier for every two digitlines. The superior signal-to-noise performance [9] of folded digitline architectures comes at the expense of larger, less efficient array layouts. Good signal-to-noise performance stems from the adjacency of true and complement digitlines and the capability to twist these digitline pairs. Sense amplifier layout is simplified because the array configuration is quarter pitch—that is, one sense amplifier for every four digitlines. Wordline driver layout is more difficult because the wordline pitch is effectively reduced in folded architectures.

The main objective of the new array architecture is to combine the advantages, while avoiding the disadvantages, of both folded and open digitline architectures. To meet this objective, the architecture needs to include the following features and characteristics:

- Open digitline mbit configuration
- Small $6F^2$ mbit
- Small, efficient array layout
- Folded digitline sense amplifier configuration
- Adjacent true and complement digitlines
- Twisted digitline pairs
- Relaxed wordline pitch
- High signal-to-noise ratio

An underlying goal of the new architecture is to reduce overall die size beyond that obtainable from either the folded or open digitline architectures. A second yet equally important goal is to achieve signal-to-noise performance that meets or approaches that of the folded digitline architecture.

Sec. 3.2   Design Examples: Advanced Bilevel DRAM Architecture   89

3.2.2 Bilevel Digitline Construction

A bilevel digitline architecture resulted from 256-Mbit DRAM research and design activities carried out at Micron Technology, Inc., in Boise, Idaho. This bilevel digitline architecture is an innovation that evolved from a comparative analysis of open and folded digitline architectures. The analysis served as a design catalyst, ultimately leading to the creation of a new DRAM array configuration—one that allows the use of $6F^2$ mbits in an otherwise folded digitline array configuration. These memory cells are a byproduct of crosspoint-style (open digitline) array blocks. Crosspoint-style array blocks require that every wordline connect to mbit transistors on every digitline, precluding the formation of digitline pairs. Yet, digitline pairs (columns) remain an essential element in folded digitline-type operation. Digitline pairs and digitline twisting are important features that provide for good signal-to-noise performance.

The bilevel digitline architecture solves the crosspoint and digitline pair dilemma through vertical integration. In vertical integration, essentially, two open digitline crosspoint array sections are placed side by side, as seen in Figure 3.7. Digitlines in one array section are designated as true digitlines, while digitlines from the second array section are designated as complement digitlines. An additional conductor is added to the DRAM process to complete the formation of the digitline pairs. The added conductor allows digitlines from each array section to route across the other array section with both true and complement digitlines vertically aligned. At the juncture between each section, the true and complement signals are vertically twisted. With this twisting, the true digitline connects to mbits in one array section, and the complement digitline connects to mbits in the other array section. This twisting concept is illustrated in Figure 3.8.

To improve the signal-to-noise characteristics of this design, the single twist region is replaced by three twist regions, as illustrated in Figure 3.9. A benefit of adding multiple twist regions is that only half of the digitline pairs actually twist within each region, making room in each region for the twists to occur. The twist regions are equally spaced at the 25%, 50%, and 75% marks in the overall array. Assuming that even digitline pairs twist at the 50% mark, odd digitlines twist at the 25% and 75% marks. Each component of a digitline pair, true and complement, spends half of its overall length on the bottom conductor connecting to mbits and half of its length on the top conductor. This characteristic balances the capacitance and the number of mbits associated with each digitline. Furthermore, the triple twisting scheme guarantees that the noise terms are balanced for each digitline, producing excellent signal-to-noise performance.

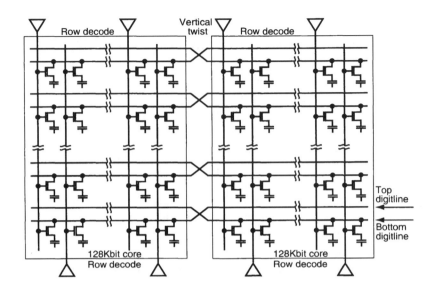

Figure 3.7 Development of bilevel digitline architecture.

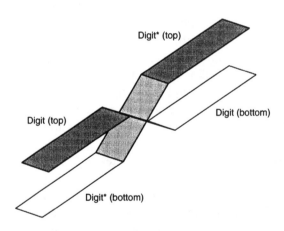

Figure 3.8 Digitline vertical twisting concept.

A variety of vertical twisting schemes is possible with the bilevel digitline architecture. As shown in Figure 3.10, each scheme uses conductive lay-

Sec. 3.2 Design Examples: Advanced Bilevel DRAM Architecture

ers already present in the DRAM process to complete the twist. Vertical twisting is simplified because only half of the digitlines are involved in a given twist region. The final selection of a twisting scheme is based on yield factors, die size, and available process technology.

To further advance the bilevel digitline architecture concept, its $6F^2$ mbit was modified to improve yield. Shown in an arrayed form in Figure 3.11, the *plaid* mbit is constructed using long parallel strips of active area vertically separated by traditional field oxide isolation. Wordlines run perpendicular to the active area in straight strips of polysilicon. Plaid mbits are again constructed in pairs that share a common contact to the digitline. Isolation gates (transistors) formed with additional polysilicon strips provide horizontal isolation between mbits. Isolation is obtained from these gates by permanently connecting the isolation gate polysilicon to either a ground or a negative potential. Using isolation gates in this mbit design eliminates one- and two-dimensional encroachment problems associated with normal isolation processes. Furthermore, many photolithography problems are eliminated from the DRAM process as a result of the straight, simple design of both the active area and the polysilicon in the mbit. The *plaid* designation for this mbit is derived from the similarity between an array of mbits and tartan fabric that is apparent in a color array plot.

In the bilevel and folded digitline architectures, both true and complement digitlines exist in the same array core. Accordingly, the sense amplifier block needs only one sense amplifier for every two digitline pairs. For the folded digitline architecture, this yields one sense amplifier for every four Metal1 digitlines—quarter pitch. The bilevel digitline architecture that uses vertical digitline stacking needs one sense amplifier for every two Metal1 digitlines—half pitch. Sense amplifier layout is therefore more difficult for bilevel than for folded designs. The three-metal DRAM process needed for bilevel architectures concurrently enables and simplifies sense amplifier layout. Metal1 is used for lower level digitlines and local routing within the sense amplifiers and row decodes. Metal2 is available for upper level digitlines and column select signal routing through the sense amplifiers. Metal3 can therefore be used for column select routing across the arrays and for control and power routing through the sense amplifiers. The function of Metal2 and Metal3 can easily be swapped in the sense amplifier block depending on layout preferences and design objectives.

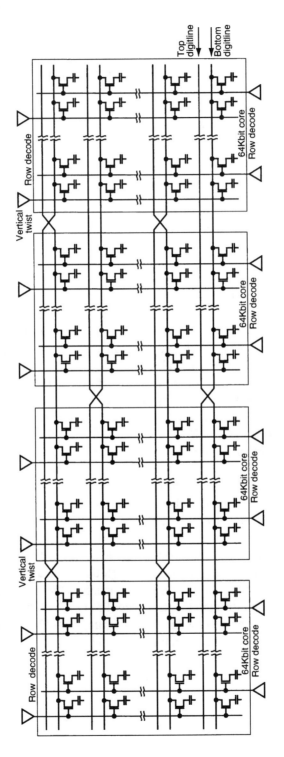

Figure 3.9 Bilevel digitline architecture schematic.

Sec. 3.2 Design Examples: Advanced Bilevel DRAM Architecture

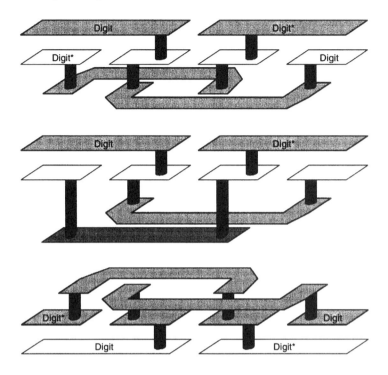

Figure 3.10 Vertical twisting schemes.

Wordline pitch is effectively relaxed for the plaid $6F^2$ mbit of the bilevel digitline architecture. The mbit is still built using the minimum process feature size of 0.3 μm. The relaxed wordline pitch stems from structural differences between a folded digitline mbit and an open digitline or plaid mbit. There are essentially four wordlines running across each folded digitline mbit pair compared to two wordlines running across each open digitline or plaid mbit pair. Although the plaid mbit is 25% shorter than a folded mbit (three versus four features), it also has half as many wordlines, effectively reducing the wordline pitch. This relaxed wordline pitch makes layout of the wordline drivers and the address decode tree much easier. In fact, both odd and even wordlines can be driven from the same row decoder block, thus eliminating half of the row decoder strips in a given array block. This is an important distinction, as the tight wordline pitch for folded digitline designs necessitates separate odd and even row decode strips.

3.2.3 Bilevel Digitline Array Architecture

The bilevel digitline array architecture depicted in Figure 3.12 is a potential architecture for tomorrow's large-scale DRAM designs. The bilevel architecture is constructed with multiple array cores separated by strips of sense amplifiers and either row decode blocks or wordline stitching regions. Wordline stitching requires a four-metal process, while row decode blocks can be implemented in a three-metal process. The array cores include 262,144 (2^{18}) functionally addressable plaid $6F^2$ mbits arranged in 532 rows and 524 bilevel digitline pairs. The 532 rows comprise 512 actual wordlines, 4 redundant wordlines, and 16 dummy wordlines. There are also 267 isolation gates in each array due to the use of plaid mbits. Because they are accounted for in the wordline pitch, however, they can be ignored. The 524 bilevel digitline pairs comprise 512 actual digitline pairs, 8 redundant digitline pairs, and 4 dummy digitline pairs. The term *digitline pair* describes the array core structure because pairing is a natural product of the bilevel architecture. Each digitline pair consists of one digitline on Metal1 and a vertically aligned complementary digitline on Metal2.

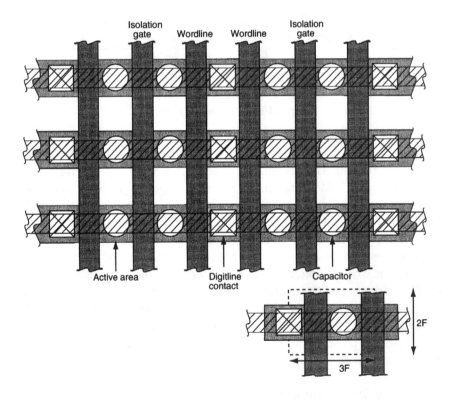

Figure 3.11 Plaid $6F^2$ mbit array.
(The isolation gates are tied to ground or a negative voltage.)

Sec. 3.2 Design Examples: Advanced Bilevel DRAM Architecture

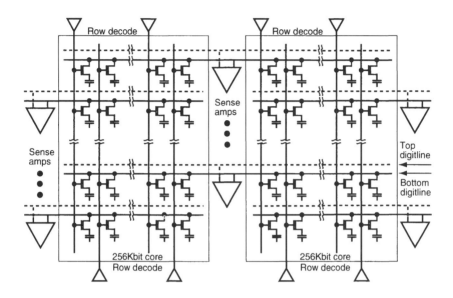

Figure 3.12 Bilevel digitline array schematic.

Sense amplifier blocks are placed on both sides of each array core. The sense amplifiers within each block are laid out at half pitch: one sense amplifier for every two Metal1 digitlines. Each sense amplifier connects through isolation devices to columns (digitline pairs) from two adjacent array cores. Similar to the folded digitline architecture, odd columns connect on one side of the array core, and even columns connect on the other side. Each sense amplifier block is then exclusively connected to either odd or even columns, never to both.

Unlike a folded digitline architecture that uses a local row decode block connected to both sides of an array core, the bilevel digitline architecture uses a local row decode block connected to only one side of each core. As stated earlier, both odd and even rows can be driven from the same local row decoder block with the relaxed wordline pitch. Because of this feature, the bilevel digitline architecture is more efficient than alternative architectures. A four-metal DRAM process allows local row decodes to be replaced by either stitch regions or local wordline drivers. Either approach could substantially reduce die size. The array core also includes the three twist regions necessary for the bilevel digitline architecture. The twist region is larger than that used in the folded digitline architecture, owing to the complexity of twisting digitlines vertically. The twist regions again constitute a break in the array structure, making it necessary to include dummy wordlines.

As with the open digitline and folded digitline architectures, the bilevel digitline length is limited by power dissipation and a minimum cell-to-digitline capacitance ratio. In the 256-Mbit generation, the digitlines are again restricted from connecting to more than 256 mbits (128 mbit pairs). The analysis to arrive at this quantity is the same as that for the open digitline architecture, except that the overall digitline capacitance is higher. The bilevel digitline runs over twice as many cells as the open digitline with the digitline running in equal lengths in both Metal2 and Metal1. The capacitance added by the Metal2 component is small compared to the already present Metal1 component because Metal2 does not connect to mbit transistors. Overall, the digitline capacitance increases by about 25% compared to an open digitline. The power dissipated during a Read or Refresh operation is proportional to the *digitline capacitance* (C_D), the supply (internal) voltage (V_{CC}), the external voltage (V_{CCX}), the number of active columns (N), and the Refresh period (P). It is given as

$$P_D = \frac{V_{CCX} \cdot (N \cdot V_{CC} \cdot (C_D + C_C))}{2 \cdot P}.$$

On a 256-Mbit DRAM in 8k Refresh, there are 32,768 (2^{15}) active columns during each Read, Write, or Refresh operation. Active array current and power dissipation for a 256-Mbit DRAM are given in Table 3.11 for a 90 ns Refresh period (–5 timing) at various digitline lengths. The budget for active array current is limited to 200mA for this 256-Mbit design. To meet this budget, the digitline cannot exceed a length of 256 mbits.

Wordline length is again limited by the maximum allowable *(RC)* time constant of the wordline. The calculation for bilevel digitline is identical to that performed for open digitline due to the similarity of array core design. These results are given in Table 3.3. Accordingly, if the wordline time constant stant is to remain under the required 4-nanosecond limit, the wordline length cannot exceed 512 mbits (512 bilevel digitline pairs).

Table 3.11 Active current and power versus bilevel digitline length.

Digitline Length (mbits)	Digitline Capacitance (fF)	Active Current	Power Dissipation (mW)
128	128	75	249
256	256	151	498
512	513	301	994

Sec. 3.2 Design Examples: Advanced Bilevel DRAM Architecture   97

A layout of various bilevel elements was generated to obtain reasonable estimates of pitch cell size. With these size estimates, overall dimensions for a 32-Mbit array block could be calculated. The diagram for a 32-Mbit array block using the bilevel digitline architecture is shown in Figure 3.13. This block requires a total of one hundred twenty-eight 256kbit array cores. The 128 array cores are arranged in 16 rows and 8 columns. Each 4-Mbit vertical section consists of 512 wordlines and 8,192 bilevel digitline pairs (8,192 columns). Eight 4-Mbit strips are required to form the complete 32-Mbit block. Sense amplifier blocks are positioned vertically between each 4-Mbit section.

Row decode strips are positioned horizontally between every array core. Only eight row decode strips are needed for the sixteen array cores, for each row decode contains wordline drivers for both odd and even rows. The 32-Mbit array block shown in Figure 3.13 includes pitch cell layout estimates. Overall size for the 32-Mbit block is found by summing the dimensions for each component.

As before,

$$Height32 = (T_R \cdot H_{RDEC}) + (T_{DL} \cdot P_{DL}) \; microns,$$

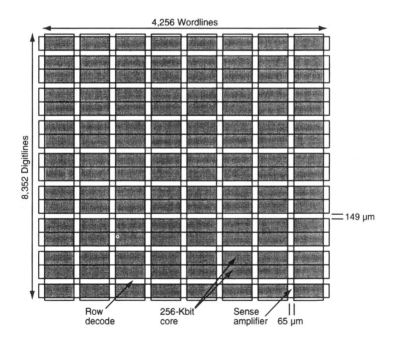

Figure 3.13 Bilevel digitline architecture 32-Mbit array block.

where T_R is the number of bilevel row decoders, H_{RDEC} is the height of each decoder, T_{DL} is the number of bilevel digitline pairs including redundant and dummy, and P_{DL} is the digitline pitch. Also,

$$Height32 = (T_{SA} \cdot W_{AMP}) + (T_{WL} \cdot P_{WL6}) + (T_{TWIST} \cdot W_{TWIST}) \; microns,$$

where T_{SA} is the number of sense amplifier strips, W_{AMP} is the width of the sense amplifiers, T_{WL} is the total number of wordlines including redundant and dummy, P_{WL6} is the wordline pitch for the plaid 6F² mbit, T_{TWIST} is the total number of twist regions, and W_{TWIST} is the width of the twist regions. Table 3.12 shows the calculated results for the bilevel 32-Mbit block shown in Figure 3.13. A three-metal process is assumed in these calculations because it requires the local row decoders. Array efficiency for the bilevel digitline 32-Mbit array block, which yields 63.1% for this design example, is given as

$$Efficiency = \frac{(100 \cdot 2^{25} \cdot P_{DL} \cdot 2 \cdot P_{WL6})}{Area32} \; percent.$$

With Metal4 added to the bilevel DRAM process, the local row decoder scheme can be replaced by a global or hierarchical row decoder scheme. The addition of a fourth metal to the DRAM process places even greater demands on process engineers. Regardless, an analysis of 32-Mbit array block size was performed assuming the availability of Metal4. The results of the analysis are shown in Tables 3.13 and 3.14 for the global and hierarchical row decode schemes. Array efficiency for the 32-Mbit memory block using global and hierarchical row decoding is 74.5% and 72.5%, respectively.

3.2.4 Architectural Comparison

Although a straight comparison of DRAM architectures might appear simple, in fact it is very complicated. Profit remains the critical test of architectural efficiency and is the true basis for comparison. This in turn requires accurate yield and cost estimates for each alternative. Without these estimates and a thorough understanding of process capabilities, conclusions are elusive and the exercise is academic. The data necessary to perform the analysis and render a decision also varies from manufacturer to manufacturer. Accordingly, a conclusive comparison of the various array architectures is beyond the scope of this book. Rather, the architectures are compared in light of the available data. To better facilitate a comparison, the 32-Mbit array block size data from Sections 3.1 and 3.2 is summarized in Table 3.15 for the open digitline, folded digitline, and bilevel digitline architectures.

Sec. 3.2 Design Examples: Advanced Bilevel DRAM Architecture 99

Table 3.12 Bilevel digitline (local row decode)—32-Mbit size calculations.

Description	Parameter	Size
Number of sense amplifier strips	T_{SA}	9
Width of sense amplifiers	W_{AMP}	65 µm
Number of local decode strips	T_{LDEC}	8
Height of local decode strips	H_{LDEC}	149 µm
Number of digitlines	T_{DL}	8,352
Number of wordlines	T_{WL}	4,256
Number of twist regions	W_{TWIST1}	9 µm
Width of twist regions	W_{TWIST}	9 µm
Height of 32-Mbit block	$Height32$	6,203 µm
Width of 32-Mbit block	$Width32$	4,632 µm
Area of 32-Mbit block	$Area32$	28,732,296 µm²

From Table 3.15, it can be concluded that overall die size (32-Mbit area) is a better metric for comparison than array efficiency. For instance, the three-metal folded digitline design using hierarchical row decodes has an area of 34,089,440mm² and an efficiency of 70.9%. The three-metal bilevel digitline design with local row decodes has an efficiency of only 63.1% but an overall area of 28,732,296 mm². Array efficiency for the folded digitline is higher. This is misleading, however, because the folded digitline yields a die that is 18.6% larger for the same number of conductors.

Table 3.13 Bilevel digitline (global decode)—32-Mbit size calculations.

Description	Parameter	Size
Number of sense amplifier strips	T_{SA}	9
Width of sense amplifiers	W_{AMP}	65 µm
Number of global decode strips	T_{GDEC}	1
Height of global decode strips	H_{GDEC}	200 µm
Number of stitch regions	N_{ST}	4
Height of stitch regions	H_{ST}	10 µm
Number of digitlines	T_{DL}	8,352
Number of wordlines	T_{WL}	4,256
Number of twist regions	T_{TWIST}	24
Width of twist regions	W_{TWIST}	9 µm
Height of 32-Mbit block	$Height32$	5,251 µm
Width of 32-Mbit block	$Width32$	4,632 µm
Area of 32-Mbit block	$Area32$	24,322,632 µm²

Table 3.15 also illustrates that the bilevel digitline architecture always yields the smallest die area, regardless of the configuration. The smallest folded digitline design at 32,654,160mm² and the smallest open digitline design at 29,944,350mm² are still larger than the largest bilevel digitline design at 28,732,296mm². It is also apparent that both the bilevel and open

Sec. 3.2 Design Examples: Advanced Bilevel DRAM Architecture 101

digitline architectures need at least three conductors in their construction. The folded digitline architecture still has a viable design option using only two conductors. The penalty of using two conductors is a much larger die size—a full 41% larger than the three-metal bilevel digitline design.

Table 3.14 Bilevel digitline (hierarchical row decode)—32-Mbit size calculations.

Description	Parameter	Size
Number of sense amplifier strips	T_{SA}	9
Width of sense amplifiers	W_{AMP}	65 µm
Number of global decode strips	T_{GDEC}	1
Height of global decode strips	H_{GDEC}	190 µm
Number of hier decode strips	N_{HDEC}	4
Height of hier decode strips	H_{HDEC}	48 µm
Number of digitlines	T_{DL}	8,352
Number of wordlines	T_{WL}	4,256
Number of twist regions	T_{TWIST}	24
Width of twist regions	W_{TWIST}	9 µm
Height of 32-Mbit block	$Height32$	5,393 µm
Width of 32-Mbit block	$Width32$	4,632 µm
Area of 32-Mbit block	$Area32$	24,930,376 µm²

Table 3.15 32-Mbit size calculations summary.

Architecture	Row Decode	Metals	32-Mbit Area (μm^2)	Efficiency (%)
Open digit	Global	3	29,944,350	60.5
Open digit	Hier	3	32,429,565	55.9
Folded digit	Local	2	40,606,720	59.5
Folded digit	Global	3	32,654,160	74.0
Folded digit	Hier	3	34,089,440	70.9
Bilevel digit	Local	3	28,732,296	63.1
Bilevel digit	Global	4	24,322,632	74.5
Bilevel digit	Hier	4	24,980,376	72.5

REFERENCES

[1] H. Hidaka, Y. Matsuda, and K. Fujishima, "A Divided/Shared Bit-Line Sensing Scheme for ULSI DRAM Cores," *IEEE Journal of Solid-State Circuits*, vol. 26, pp. 473–477, April 1991.

[2] T. Hamada, N. Tanabe, H. Watanabe, K. Takeuchi, N. Kasai, H. Hada, K. Shibahara, K. Tokashiki, K. Nakajima, S. Hirasawa, E. Ikawa, T. Saeki, E. Kakehashi, S. Ohya, and T. A. Kunio, "A Split-Level Diagonal Bit-Line (SLDB) Stacked Capacitor Cell for 256Mb Drams," *1992 IEDM Technical Digest*, pp. 799–802.

[3] M. Inoue, H. Kotani, T. Yamada, H. Yamauchi, A. Fujiwara, J. Matsushima, H. Akamatsu, M. Fukumoto, M. Kubota, I. Nakao, N. Aoi, G. Fuse, S. Ogawa, S. Odanaka, A. Ueno, and H. Yamamoto, "A 16Mb DRAM with an Open Bit-Line Architecture," *1988 IEEE ISSCC Digest of Technical Papers*, pp. 246–247.

[4] M. Inoue, T. Yamada, H. Kotani, H. Yamauchi, A. Fujiwara, J. Matsushima, H. Akamatsu, M. Fukumoto, M. Kubota, I. Nakao, N. Aoi, G. Fuse, S. Ogawa, S. Odanaka, A. Ueno, and H. Yamamoto, "A 16-Mbit DRAM with a Relaxed Sense-Amplifier-Pitch Open-Bit-Line Architecture," *IEEE Journal of Solid-State Circuits*, vol. 23, pp. 1104–1112, October 1988.

[5] K. Noda, T. Saeki, A. Tsujimoto, T. Murotani, and K. Koyama, "A Boosted Dual Word-line Decoding Scheme for 256Mb DRAMs," *1992 Symposium on VLSI Circuits Digest of Technical Papers*, pp. 112–113.

References

[6] D. Takashima, S. Watanabe, H. Nakano, Y. Oowaki, and K. Ohuchi, "Open/Folded Bit-Line Arrangement for Ultra-High-Density DRAM's," *IEEE Journal of Solid-State Circuits,* vol. 29, pp. 539–542, April 1994.

[7] Y. Oowaki, K. Tsuchida, Y. Watanabe, D. Takashima, M. Ohta, H. Nakano, S. Watanabe, A. Nitayama, F. Horiguchi, K. Ohuchi, and F. Masuoka, "A 33-ns 64Mb DRAM," *IEEE Journal of Solid-State Circuits,* vol. 26, pp. 1498–1505, November 1991.

[8] Y. Nakagome, M. Aoki, S. Ikenaga, M. Horiguchi, S. Kimura, Y. Kawamoto, and K. Itoh, "The Impact of Data-Line Interference Noise on DRAM Scaling," *IEEE Journal of Solid-State Circuits,* vol. 23, pp. 1120–1127, October 1988.

[9] T. Yoshihara, H. Hidaka, Y. Matsuda, and K. Fujishima, "A Twisted Bitline Technique for Multi-Mb DRAMs," *1988 IEEE ISSCC Digest of Technical Papers,* pp. 238–239.

Chapter 4

The Peripheral Circuitry

In this chapter, we briefly discuss the peripheral circuitry. In particular, we discuss the column decoder and its implementation. We also cover the implementation of row and column redundancy.

4.1 COLUMN DECODER ELEMENTS

The column decoder circuits represent the final DRAM elements that pitch up to the array mbits. Historically, column decode circuits were simple and straightforward: static logic gates were generally used for both the decode tree elements and the driver output. Static logic was used primarily because of the nature of column addressing in *fast page mode* (FPM) and *extended data out* (EDO) devices. Unlike row addressing, which occurred once per $\overline{RAS}$ cycle, column addressing could occur multiple times per $\overline{RAS}$ cycle, with each column held open until a subsequent column address appeared. The transition interval from one column to the next had to be minimized, allowing just enough time to turn OFF the previous column, turn ON the new column, and equilibrate the necessary data path circuits. Furthermore, because column address transitions were unpredictable and asynchronous, a column decode logic that was somewhat forgiving of random address changes was required—hence the use of static logic gates.

With the advent of *synchronous DRAMs* (SDRAMs) and high-speed, packet-based DRAM technology, the application of column, or row, addresses became synchronized to a clock. More importantly, column addressing and column timing became more predictable, allowing design engineers to use pipelining techniques along with dynamic logic gates in constructing column decode elements.

Column redundancy adds complexity to the column decoder because the redundancy operation in FPM and EDO DRAMs requires the decode circuit to terminate column transitions, prior to completion. In this way, redundant column elements can replace normal column elements. Generally, the addressed column select is allowed to fire normally. If a redundant column match occurs for this address, the normal column select is subsequently turned OFF; the redundant column select is fired. The redundant match is timed to disable the addressed column select before enabling the I/O devices in the sense amplifier.

The fire-and-cancel operation used on the FPM and EDO column decoders is best achieved with static logic gates. In packet-based and synchronous DRAMs, column select firing can be synchronized to the clock. Synchronous operation, however, does not favor a fire-and-cancel mode, preferring instead that the redundant match be determined prior to firing either the addressed or redundant column select. This match is easily achieved in a pipeline architecture because the redundancy match analysis can be performed upstream in the address pipeline before presenting the address to the column decode logic.

A typical FPM- or EDO-type column decoder realized with static logic gates is shown schematically in Figure 4.1. The address tree is composed of combinations of NAND or NOR gates. In this figure, the address signals are active HIGH, so the tree begins with two-input NAND gates. Using predecoded address lines is again preferred. Predecoded address lines both simplify and reduce the decoder logic because a single input can represent two or more address terms. In the circuit shown in Figure 4.1, the four input signals *CA*23, *CA*45, *CA*67, and *CA*8 represent seven address terms, permitting 1 of 128 decoding. Timing of the column selection is controlled by a signal called *column decode enable (CDE)*, which is usually combined with an input signal, as shown in Figure 4.2, or as an additional term in the tree.

The final signal shown in the figure is labeled *RED*. This signal disables the normal column term and enables redundant column decoders whenever the column address matches any of the redundant address locations, as determined by the column-redundant circuitry. A normal column select begins to turn ON before *RED* makes a LOW-to-HIGH transition. This fire-and-cancel operation maximizes the DRAM speed between column accesses, making column timing all the more critical. An example, which depicts a page mode column transition, is shown in Figure 4.2. During the fire-and-cancel transition, *I/O equilibration (EQIO)* envelops the deselection of the old column select *CSEL*<0> and the selection of a new column select *RCSEL*<1>. In this example, *CSEL*<1> initially begins to turn ON until *RED* transitions HIGH, disabling *CSEL*<1> and enabling redundant column *RCSEL*<1>.

Sec. 4.1 Column Decoder Elements

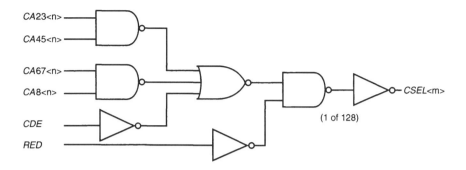

Figure 4.1 Column decode.

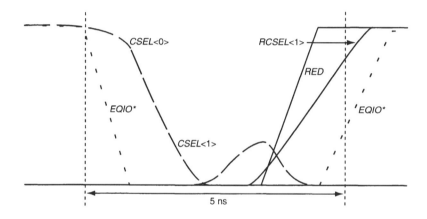

Figure 4.2 Column selection timing.

The column decode output driver is a simple CMOS inverter because the *column select signal (CSEL)* only needs to drive to V_{CC}. On the other hand, the wordline driver, as we have seen, is rather complex; it needed to drive to a boosted voltage, V_{CCP}. Another feature of column decoders is that their pitch is very relaxed relative to the pitch of the sense amplifiers and row decoders. From our discussion in Section 2.2 concerning I/O transistors and *CSEL* lines, we learned that a given *CSEL* is shared by four to eight I/O transistors. Therefore, the *CSEL* pitch is one *CSEL* for every eight to sixteen digitlines. As a result, the column decoder layout is much less difficult to implement than either the row decoder or the sense amplifiers.

A second type of column decoder, realized with dynamic P&E logic, is shown in Figure 4.3. This particular design was first implemented in an 800MB/s packet-based SLDRAM device. The packet nature of SLDRAM and the extensive use of pipelining supported a column decoder built with P&E logic. The column address pipeline included redundancy match circuits upstream from the actual column decoder, so that both the column address and the corresponding match data could be presented at the same time. There was no need for the fire-and-cancel operation: the match data was already available.

Therefore, the column decoder fires either the addressed column select or the redundant column select in synchrony with the clock. The decode tree is similar to that used for the CMOS wordline driver; a pass transistor was added so that a decoder enable term could be included. This term allows the tree to disconnect from the latching column select driver while new address terms flow into the decoder. A latching driver was used in this pipeline implementation because it held the previously addressed column select active with the decode tree disconnected. Essentially, the tree would disconnect after a column select was fired, and the new address would flow into the tree in anticipation of the next column select. Concurrently, redundant match information would flow into the phase term driver along with $CA45$ address terms to select the correct phase signal. A redundant match would then override the normal phase term and enable a redundant phase term.

Operation of this column decoder is shown in Figure 4.4. Once again, deselection of the old column select $CSEL<0>$ and selection of a new column select $RCSEL<1>$ are enveloped by $EQIO$. Column transition timing is under the control of the column latch signal $CLATCH*$. This signal shuts OFF the old column select and enables firing of the new column select. Concurrent with $CLATCH*$ firing, the decoder is enabled with *decoder enable (DECEN)* to reconnect the decode tree to the column select driver. After the new column select fires, $DECEN$ transitions LOW to once again isolate the decode tree.

4.2 COLUMN AND ROW REDUNDANCY

Redundancy has been used in DRAM designs since the 256k generation to improve yield and profitability. In redundancy, spare elements such as rows and columns are used as logical substitutes for defective elements. The substitution is controlled by a physical encoding scheme. As memory density and size increase, redundancy continues to gain importance. The early designs might have used just one form of repairable elements, relying exclusively on row or column redundancy. Yet as processing complexity increased and feature size shrank, both types of redundancy—row and column—became mandatory.

Sec. 4.2 Column and Row Redundancy

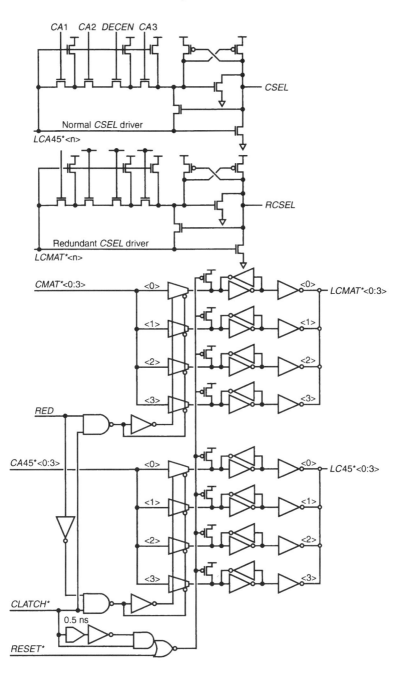

Figure 4.3 Column decode: P&E logic.

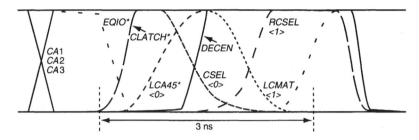

Figure 4.4 Column decode waveforms.

Today various DRAM manufacturers are experimenting with additional forms of repair, including replacing entire subarrays. The most advanced type of repair, however, involves using movable saw lines as realized on a prototype 1Gb design [1]. Essentially, any four good adjacent quadrants from otherwise bad die locations can be combined into a good die by simply sawing the die along different scribe lines. Although this idea is far from reaching production, it illustrates the growing importance of redundancy.

4.2.1 Row Redundancy

The concept of row redundancy involves replacing bad wordlines with good wordlines. There could be any number of problems on the row to be repaired, including shorted or open wordlines, wordline-to-digitline shorts, or bad mbit transistors and storage capacitors. The row is not physically but logically replaced. In essence, whenever a row address is strobed into a DRAM by $\overline{RAS}$, the address is compared to the addresses of known bad rows. If the address comparison produces a match, then a replacement wordline is fired in place of the normal (bad) wordline.

The replacement wordline can reside anywhere on the DRAM. Its location is not restricted to the array containing the normal wordline, although its range may be restricted by architectural considerations. In general, the redundancy is considered local if the redundant wordline and the normal wordline must always be in the same subarray.

If, however, the redundant wordline can exist in a subarray that does not contain the normal wordline, the redundancy is considered global. Global repair generally results in higher yield because the number of rows that can be repaired in a single subarray is not limited to the number of its redundant rows. Rather, global repair is limited only by the number of fuse banks, termed *repair elements,* that are available to any subarray.

Local row repair was prevalent through the 16-Meg generation, producing adequate yield for minimal cost. Global row repair schemes are becom-

Sec. 4.2 Column and Row Redundancy   111

ing more common for 64-Meg or greater generations throughout the industry. Global repair is especially effective for repairing clustered failures and offers superior repair solutions on large DRAMs.

Dynamic logic is a traditional favorite among DRAM designers for row redundant match circuits. Dynamic gates are generally much faster than static gates and well suited to row redundancy because they are used only once in an entire $\overline{RAS}$ cycle operation. The dynamic logic we are referring to again is called *precharge and evaluate* (P&E). Match circuits can take many forms; a typical row match circuit is shown in Figure 4.5. It consists of a *PRECHARGE* transistor M1, match transistors M2–M5, laser fuses F1–F4, and static gate I1. In addition, the node labeled *row PRECHARGE (RPRE*)* is driven by static logic gates. The fuses generally consist of narrow polysilicon lines that can be blown or opened with either a precision laser or a fusing current provided by additional circuits (not shown).

For our example using predecoded addresses, three of the four fuses shown must be blown in order to program a match address. If, for instance, F2–F4 were blown, the circuit would match for *RA*12<0> but not for *RA*12<1:3>. Prior to $\overline{RAS}$ falling, the node labeled *RED** is precharged to V_{CC} by the signal *RPRE**, which is LOW. Assuming that the circuit is enabled by fuse F5, *EVALUATE** will be LOW. After $\overline{RAS}$ falls, the row addresses eventually propagate into the redundant block. If *RA*12<0> fires HIGH, *RED** discharges through M2 to ground. If, however, *RA*12<0> does not fire HIGH, *RED** remains at V_{CC}, indicating that a match did not occur. A weak latch composed of I1 and M6 ensures that *RED** remains at V_{CC} and does not discharge due to junction leakage.

This latch can easily be overcome by any of the match transistors. The signal *RED** is combined with static logic gates that have similar *RED** signals derived from the remaining predecoded addresses. If all of the *RED** signals for a redundant element go LOW, then a match has occurred, as indicated by *row match (RMAT*)* firing LOW. The signal *RMAT** stops the normal row from firing and selects the appropriate replacement wordline. The fuse block in Figure 4.5 shows additional fuses F5–F6 for enabling and disabling the fuse bank. Disable fuses are important in the event that the redundant element fails and the redundant wordline must itself be repaired.

The capability to pretest redundant wordlines is an important element in most DRAM designs today. For the schematic shown in Figure 4.5, the pretest is accomplished through the set of static gates driven by the *redundant test (REDTEST)* signal. The input signals labeled *TRA*12n, *TRA*34n, *TRA*56n, and *TODDEVEN* are programmed uniquely for each redundant element through connections to the appropriate predecoded address lines.

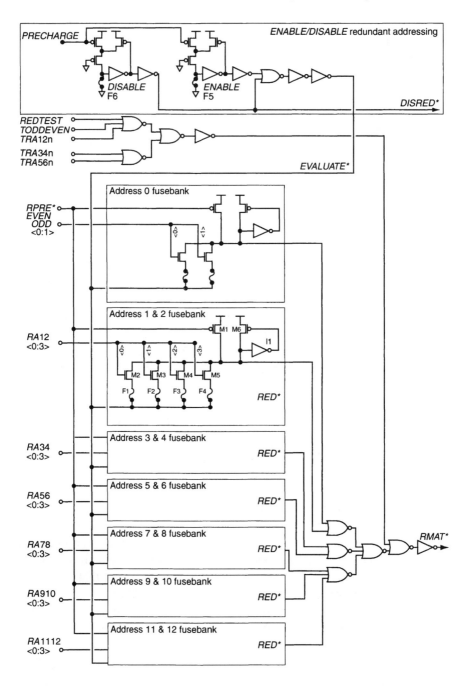

Figure 4.5 Row fuse block.

Sec. 4.2 Column and Row Redundancy 113

REDTEST is HIGH whenever the DRAM is in the redundant row pretest mode. If the current row address corresponds to the programmed pretest address, *RMAT** will be forced LOW, and the corresponding redundant wordline rather than the normal wordline will be fired. This pretest capability permits all of the redundant wordlines to be tested prior to any laser programming.

Fuse banks or redundant elements, as shown in Figure 4.5, are physically associated with specific redundant wordlines in the array. Each element can fire only one specific wordline, although generally in multiple subarrays. The number of subarrays that each element controls depends on the DRAM's architecture, refresh rate, and redundancy scheme. It is not uncommon in 16-Meg DRAMs for a redundant row to replace physical rows in eight separate subarrays at the same time. Obviously, the match circuits must be fast. Generally, firing of the normal row must be held off until the match circuits have enough time to evaluate the new row address. As a result, time wasted during this phase shows up directly on the part's *row access (tRAC)* specification.

4.2.2 Column Redundancy

Column redundancy is the second type of repair available on most DRAM designs. In Section 4.1, it was stated that column accesses can occur multiple times per $\overline{RAS}$ cycle. Each column is held open until a subsequent column appears. Therefore, column redundancy was generally implemented with circuits that are very different from those seen in row redundancy. As shown in Figure 4.6, a typical column fuse block is built from static logic gates rather than from P&E dynamic gates. P&E logic, though extremely fast, needs to be precharged prior to each evaluation. The transition from one column to another must occur within the span of a few nanoseconds. On FPM and EDO devices, which had unpredictable and asynchronous column address transitions, there was no guarantee that adequate time existed for this *PRECHARGE* to occur. Yet the predictable nature of column addressing on the newer type of synchronous and packet-based DRAMs affords today's designers an opportunity to use P&E-type column redundancy circuits. In these cases, the column redundant circuits appear very similar to P&E-style row redundant circuits.

An example of a column fuse block for an FPM or EDO design is shown in Figure 4.6. This column fuse block has four sets of column fuse circuits and additional enable/disable logic. Each column fuse circuit, corresponding to a set of predecoded column addresses, contains compare logic and two sets of laser fuse/latch circuits. The laser fuse/latch reads the laser fuse whenever *column fuse power (CFP)* is enabled, generally on POWERUP and during $\overline{RAS}$ cycles. The fuse values are held by the simple inverter latch circuits composed of I0 and I1. Both true and complement data are fed from the fuse/latch circuit into the comparator logic. The comparator logic, which appears somewhat complex, is actually quite simple as shown in the following Boolean expression where $F0$ without the bar indicates a blown fuse:

$$CAM = (\overline{A0} \bullet \overline{F0} \bullet \overline{F1}) + (\overline{A1} \bullet F0 \bullet \overline{F1}) + (\overline{A2} \bullet \overline{F0} \bullet F1) + (\overline{A3} \bullet F0 \bullet F1).$$

The *column address match (CAM)* signals from all of the predecoded addresses are combined in standard static logic gates to create a *column match (CMAT*)* signal for the column fuse block. The *CMAT** signal, when active, cancels normal *CSEL* signals and enables redundant *RCSEL* signals, as described in Section 4.1. Each column fuse block is active only when its corresponding enable fuse has been blown. The column fuse block usually contains a disable fuse for the same reason as a row redundant block: to repair a redundant element. Column redundant pretest is implemented somewhat differently in Figure 4.6 than row redundant pretest here. In Figure 4.6, the bottom fuse terminal is not connected directly to ground. Rather, all of the signals for the entire column fuse block are brought out and programmed either to ground or to a column pretest signal from the test circuitry.

During standard part operation, the pretest signal is biased to ground, allowing the fuses to be read normally. However, during column redundant pretest, this signal is brought to V_{CC}, which makes the laser fuses appear to be programmed. The fuse/latch circuits latch the apparent fuse states on the next $\overline{RAS}$ cycle. Then, subsequent column accesses allow the redundant column elements to be pretested by merely addressing them via their pre-programmed match addresses.

The method of pretesting just described always uses the match circuits to select a redundant column. It is a superior method to that described for the row redundant pretest because it tests both the redundant element and its match circuit. Furthermore, as the match circuit is essentially unaltered during redundant column pretest, the test is a better measure of the obtainable DRAM performance when the redundant element is active.

Obviously, the row and column redundant circuits that are described in this section are only one embodiment of what could be considered a wealth of possibilities. It seems that all DRAM designs use some alternate form of redundancy. Other types of fuse elements could be used in place of the laser fuses that are described. A simple transistor could replace the laser fuses in either Figure 4.5 or Figure 4.6, its gate being connected to an alternative fuse element. Furthermore, circuit polarity could be reversed and non-predecoded addressing and other types of logic could be used. The options are nearly limitless. Figure 4.7 shows a SEM image of a set of poly fuses.

Sec. 4.2 Column and Row Redundancy

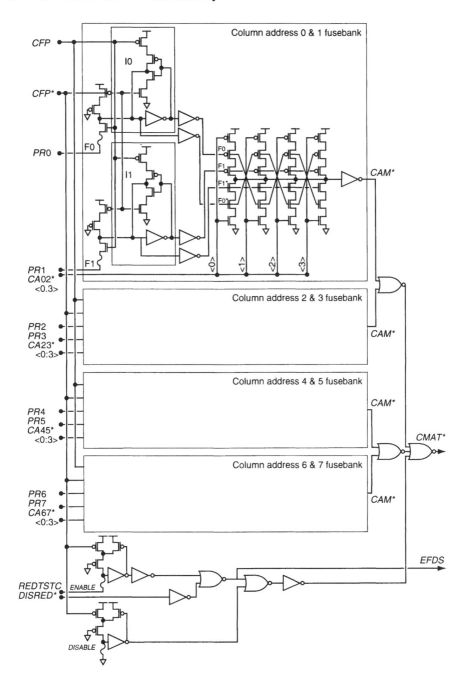

Figure 4.6 Column fuse block.

Figure 4.7 8-Meg x 8-sync DRAM poly fuses.

REFERENCE

[1] T. Sugibayashi, I. Naritake, S. Utsugi, K. Shibahara, R. Oikawa, H. Mori, S. Iwao, T. Murotani, K. Koyama, S. Fukuzawa, T. Itani, K. Kasama, T. Okuda, S. Ohya, and M. Ogawa, "A 1Gbit DRAM for File Applications," *Digest of International Solid-State Circuits Conference,* pp. 254–255, 1995.

Chapter 5

Global Circuitry and Considerations

In this chapter, we discuss the circuitry and design considerations associated with the circuitry external to the DRAM memory array and memory array peripheral circuitry. We call this *global circuitry*.

5.1 DATA PATH ELEMENTS

The typical DRAM data path is bidirectional, allowing data to be both written to and read from specific memory locations. Some of the circuits involved are truly bidirectional, passing data in for Write operations and out for Read operations. Most of the circuits, however, are unidirectional, operating on data in only a Read or Write operation. To support both operations, therefore, unidirectional circuits occur in complementary pairs: one for reading and one for writing. Operating the same regardless of the data direction, sense amplifiers, I/O devices, and data muxes are examples of bidirectional circuits. Write drivers and data amplifiers, such as *direct current sense amplifiers* (DCSAs) or data input buffers and data output buffers, are examples of paired, unidirectional circuits. In this chapter, we explain the operation and design of each of these elements and then show how the elements combine to form a DRAM data path. In addition, we discuss address test compression circuits and data test compression circuits and how they affect the overall data path design.

5.1.1 Data Input Buffer

The first element of any DRAM data path is the data input buffer. Shown in Figure 5.1, the input buffer consists of both NMOS and PMOS transistors, basically forming a pair of cascaded inverters. The first inverter stage has *ENABLE* transistors M1 and M2, allowing the buffer to be powered down during inactive periods. The transistors are carefully sized to provide high-speed operation and specific input trip points. The *high-input trip point* V_{IH} is set to 2.0 V for *low-voltage TTL (LVTTL)*-compatible DRAMs, while the *low-input trip point* V_{IL} is set to 0.8 V.

Designing an input buffer to meet specified input trip points generally requires a flexible design with a variety of transistors that can be added or deleted with edits to the metal mask. This is apparent in Figure 5.1 by the presence of switches in the schematic; each switch represents a particular metal option available in the design. Because of variations in temperature, device, and process, the final input buffer design is determined with actual silicon, not simulations. For a DRAM that is 8 bits wide (x8), there will be eight input buffers, each driving into one or more Write driver circuits through a signal labeled *DW<n>* (Data Write where n corresponds to the specific data bit 0–7).

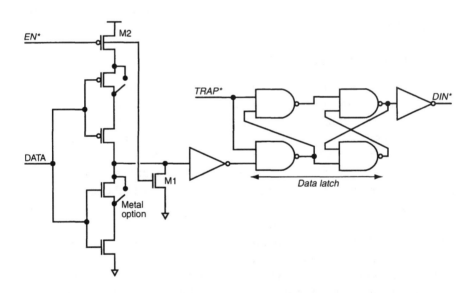

Figure 5.1 Data input buffer.

As the power supply drops, the basic inverter-based input buffer shown in Figure 5.1 is finding less use in DRAM. The required noise margins and speed of the interconnecting bus between the memory controller and the DRAM are getting difficult to meet. One high-speed bus topology, called *stub series terminated logic* (SSTL), is shown in Figure 5.2 [1]. Tightly con-

Sec. 5.1 Data Path Elements 119

trolled transmission line impedances and series resistors transmit high-speed signals with little distortion. Figure 5.2a shows the bus for clocks, command signals, and addresses. Figure 5.2b shows the bidirectional bus for transmitting data to and from the DRAM controller. In either circuit, V_{TT} and V_{REF} are set to $V_{CC}/2$.

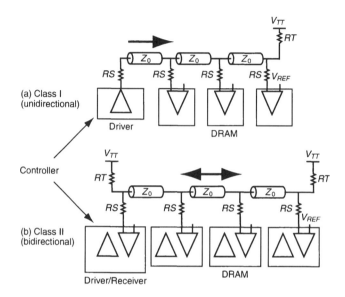

Figure 5.2 Stub series terminated logic (SSTL).

From this topology, we can see that a fully differential input buffer should be used: an inverter won't work. Some examples of fully differential input buffers are seen in Figure 5.3 [1][2], Figure 5.4 [2], and Figure 5.5 [1].

Figure 5.3 is simply a CMOS differential amplifier with an inverter output to generate valid CMOS logic levels. Common-mode noise on the diff-amp inputs is, ideally, rejected while amplifying the difference between the input signal and the reference signal. The diff-amp input common-mode range, say a few hundred mV, sets the minimum input signal amplitude (centered around V_{REF}) required to cause the output to change stages. The speed of this configuration is limited by the diff-amp biasing current. Using a large current will increase input receiver speed and, at the same time, decrease amplifier gain and reduce the diff-amp's input common-mode range.

The input buffer of Figure 5.3 requires an external biasing circuit. The circuit of Figure 5.4 is *self-biasing*. This circuit is constructed by joining a p-channel diff-amp and an n-channel diff-amp at the active load terminals.

(The active current mirror loads are removed.) This circuit is simple and, because of the adjustable biasing connection, potentially very fast. An output inverter, which is not shown, is often needed to ensure that valid output logic levels are generated.

Both of the circuits in Figures 5.3 and 5.4 suffer from duty-cycle distortion at high speeds. The PULLUP delay doesn't match the PULLDOWN delay. Duty-cycle distortion becomes more of a factor in input buffer design as synchronous DRAMs move toward clocking on both the rising and falling edges of the system clock. The fully differential self-biased input receiver of Figure 5.5 provides an adjustable bias, which acts to stabilize the PULLUP and PULLDOWN drives. An inverter pair is still needed on the output of the receiver to generate valid CMOS logic levels (two inverters in cascade on each output). A pair of inverters is used so that the delay from the inverter pairs' input to its output is constant independent of a logic one or zero propagating through the pair.

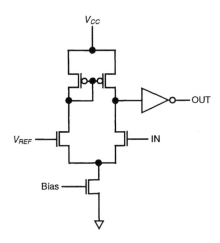

Figure 5.3 Differential amplifier-based input receiver.

Sec. 5.1 Data Path Elements 121

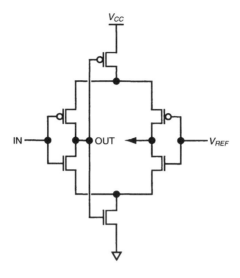

Figure 5.4 Self-biased differential amplifier-based input buffer.

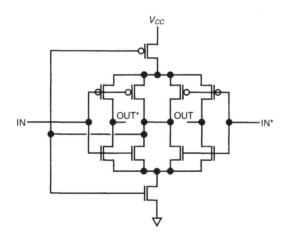

Figure 5.5 Fully differential amplifier-based input buffer.

5.1.2 Data Write Muxes

Data muxes are often used to extend the versatility of a design. Although some DRAM designs connect the input buffer directly to the Write driver circuits, most architectures place a block of Data Write muxes between the input buffers and the Write drivers. The muxes allow a given DRAM design to support multiple configurations, such as x4, x8, and x16 *I/O*. A typical schematic for these muxes is shown in Figure 5.6. As shown in this figure, the muxes are programmed according to the bond option control signals labeled *OPTX*4, *OPTX*8, and *OPTX*16. For x16 operation, each input buffer is muxed to only one set of *DW* lines. For x8 operation, each input buffer is muxed to two sets of *DW* lines, essentially doubling the quantity of mbits available to each input buffer. For x4 operation, each input buffer is muxed to four sets of *DW* lines, again doubling the number of mbits available to the remaining four operable input buffers.

Essentially, as the quantity of input buffers is reduced, the amount of column address space for the remaining buffers is increased. This concept is easy to understand as it relates to a 16Mb DRAM. As a x16 part, this DRAM has 1 mbit per data pin; as a x8 part, 2 mbits per data pin; and as a x4 part, 4 mbits per data pin. For each configuration, the number of array sections available to an input buffer must change. By using Data Write muxes that permit a given input buffer to drive as few or as many Write driver circuits as required, design flexibility is easily accommodated.

5.1.3 Write Driver Circuit

The next element in the data path to be considered is the Write driver circuit. This circuit, as the name implies, writes data from the input buffers into specific memory locations. The Write driver, as shown in Figure 5.7, drives specific *I/O* lines coming from the mbit arrays. A given Write driver is generally connected to only one set of *I/O* lines, unless multiple sets of *I/O* lines are fed by a single Write driver circuit via additional muxes. Using muxes between the Write driver and the arrays to limit the number of Write drivers and DCSA circuits is quite common. Regardless, the Write driver uses a tristate output stage to connect to the *I/O* lines. Tristate outputs are necessary because the *I/O* lines are used for both Read and Write operations. The Write driver remains in a high-impedance state unless the signal labeled Write is HIGH and either *DW* or *DW** transitions LOW from the initial HIGH state, indicating a Write operation. As shown in Figure 5.7, the Write driver is controlled by specified column addresses, the Write signal, and *DW*<n>. The driver transistors are sized large enough to ensure a quick, efficient Write operation. This is important because the array sense amplifiers usually remain ON during a Write cycle.

The remaining elements of the Write Data path reside in the array and pitch circuits. As previously discussed in Sections 1.2, 2.1, and 2.2, the mbits and sense amplifier block constitute the end of the Write Data path. The new input data is driven by the Write driver, propagating through the

Sec. 5.1 Data Path Elements

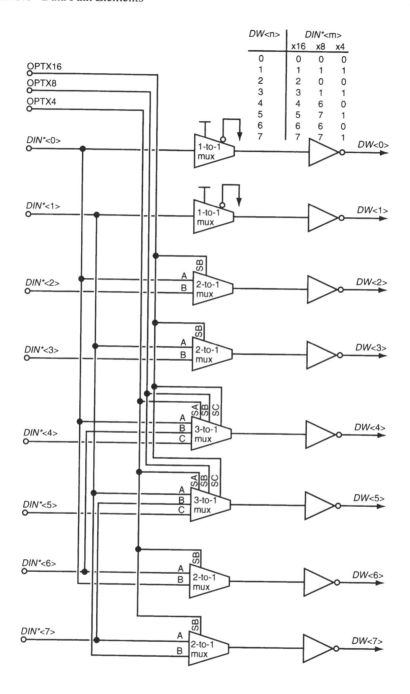

Figure 5.6 Data Write mux.

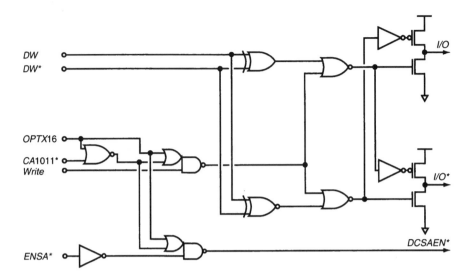

Figure 5.7 Write driver.

I/O transistors and into the sense amplifier circuits. After the sense amplifiers are overwritten and the new data is latched, the Write drivers are no longer needed and can be disabled. Completion of the Write operation into the mbits is accomplished by the sense amplifiers, which restore the digit-lines to full V_{CC} and ground levels. See Sections 1.2 and 2.2 for further discussion.

5.1.4 Data Read Path

The Data Read path is similar, yet complementary, to the Data Write path. It begins, of course, in the array, as previously discussed in Sections 1.2 and 2.2. After data is read from the mbit and latched by the sense amplifiers, it propagates through the I/O transistors onto the *I/O* signal lines and into a *DC sense amplifier* (DCSA) or *helper flip-flop* (HFF). The *I/O* lines, prior to the *column select (CSEL)* firing, are equilibrated and biased to a voltage approaching V_{CC}. The actual bias voltage is determined by the *I/O* bias circuit, which serves to control the *I/O* lines through every phase of the Read and Write cycles. This circuit, as shown in Figure 5.8, consists of a group of *bias and equilibrate* transistors that operate in concert with a variety of control signals. When the DRAM is in an idle state, such as when $\overline{RAS}$ is HIGH, the *I/O* lines are generally biased to V_{CC}. During a Read cycle and prior to *CSEL*, the bias is reduced to approximately one V_{TH} below V_{CC}.

Sec. 5.1 Data Path Elements 125

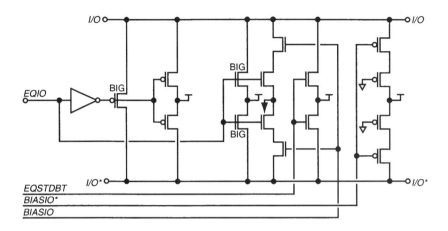

Figure 5.8 *I/O* bias circuit.

The actual bias voltage for a Read operation is optimized to ensure sense amplifier stability and fast sensing by the DCSA or HFF circuits. Bias is maintained continually throughout a Read cycle to ensure proper DCSA operation and to speed equilibration between cycles by reducing the range over which the *I/O* lines operate. Furthermore, because the *DCSA*s or *HFF*s are very high-gain amplifiers, rail-to-rail input signals are not necessary to drive the outputs to CMOS levels. In fact, it is important that the input levels not exceed the DCSA or HFF common-mode operating range. During a Write operation, the bias circuits are disabled by the Write signal, permitting the Write drivers to drive rail-to-rail.

Operation of the bias circuits is seen in the signal waveforms shown in Figure 5.9. For the Read-Modify-Write cycle, the *I/O* lines start at V_{CC} during standby; transition to $V_{CC}-V_{TH}$ at the start of a Read cycle; separate but remain biased during the Read cycle; drive rail-to-rail during a Write cycle; recover to Read cycle levels (termed *Write Recovery*); and equilibrate to $V_{CC}-V_{TH}$ in preparation for another Read cycle.

5.1.5 DC Sense Amplifier (DCSA)

The next data path element is the *DC sense amplifier* (DCSA). This amplifier, termed Data amplifier or Read amplifier by DRAM manufacturers, is an essential component in modern, high-speed DRAM designs and takes a variety of forms. In essence, the DCSA is a high-speed, high-gain differential amplifier for amplifying very small Read signals appearing on the *I/O* lines into full CMOS data signals used at the output data buffer. In most designs, the *I/O* lines connected to the sense amplifiers are very capacitive.

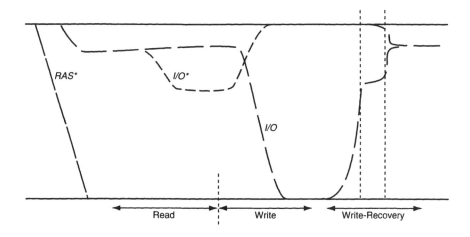

Figure 5.9 *I/O* bias and operation waveforms.

The array sense amplifiers have very limited drive capability and are unable to drive these lines quickly. Because the DCSA has very high gain, it amplifies even the slightest separation of the *I/O* lines into full CMOS levels, essentially gaining back any delay associated with the *I/O* lines. Good DCSA designs can output full rail-to-rail signals with input signals as small as 15mV. This level of performance can only be accomplished through very careful design and layout. Layout must follow good analog design principles, with each element a direct copy (no mirrored layouts) of any like elements.

As illustrated in Figure 5.10, a typical DCSA consists of four differential pair amplifiers and self-biasing CMOS stages. The differential pairs are configured as two sets of balanced amplifiers. Generally, the amplifiers are built with an NMOS differential pair using PMOS active loads and NMOS current mirrors. Because NMOS has higher mobility, providing for smaller transistors and lower parasitic loads, NMOS amplifiers usually offer faster operation than PMOS amplifiers. Furthermore, V_{TH} matching is usually better for NMOS, offering a more balanced design. The first set of amplifiers is fed with *I/O* and *I/O** signals from the array; the second set, with the output signals from the first pair, labeled *DX* and *DX**. Bias levels into each stage are carefully controlled to provide optimum performance.

The outputs from the second stage, labeled *DY* and *DY**, feed into self-biasing CMOS inverter stages for fast operation. The final output stage is capable of tristate operation to allow multiple sets of DCSA to drive a given set of *Data Read* lines *(DR<n> and DR*<n>)*. The entire amplifier is equil-

Sec. 5.1 Data Path Elements   127

ibrated prior to operation, including the self-biasing CMOS inverter stages, by all of the devices connected to the signals labeled *EQSA*, *EQSA**, and *EQSA* 2. Equilibration is necessary to ensure that the amplifier is electrically balanced and properly biased before the input signals are applied. The amplifier is enabled whenever *ENSA** is brought LOW, turning ON the output stage and the current mirror bias circuit, which is connected to the differential amplifiers via the signal labeled *CM*. For a DRAM Read cycle, operation of this amplifier is depicted in the waveforms shown in Figure 5.11. The bias levels are reduced for each amplifier stage, approaching $V_{CC}/2$ for the final stage.

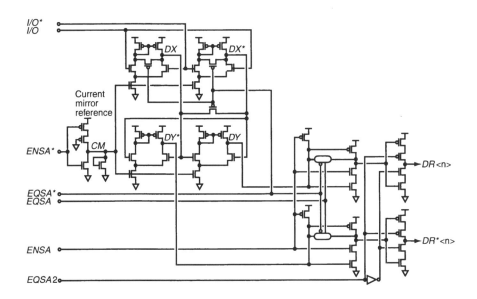

Figure 5.10 DC sense amp.

5.1.6 Helper Flip-Flop (HFF)

The DCSA of the last section can require a large layout area. To reduce the area, a helper flip-flop (HFF) as seen in Figure 5.12, can be used. The HFF is basically a clocked connection of two inverters as a latch [3]. When *CLK* is LOW, the *I/O* lines are connected to the inputs/outputs of the inverters. The inverters don't see a path to ground because M1 is OFF when *CLK* is LOW. When *CLK* transitions HIGH, the outputs of the HFF amplify, in effect, the inputs into full logic levels.

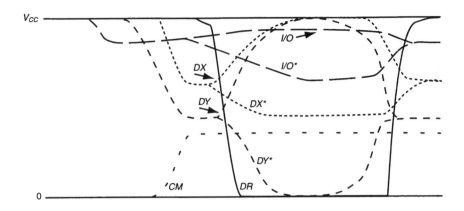

Figure 5.11 DCSA operation waveforms.

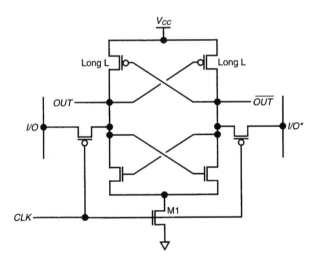

Figure 5.12 A helper flip-flop.

For example, if I/O = 1.25 V and I/O^* = 1.23 V, then I/O becomes V_{CC}, and I/O^* goes to zero when CLK transitions HIGH. Using positive feedback makes the HFF sensitive and fast. Note that HFFs can be used at several locations on the I/O lines due to the small size of the circuit.

Sec. 5.1 Data Path Elements   129

5.1.7 Data Read Muxes

The Read Data path proceeds from the DCSA block to the output buffers. The connection between these elements can either be direct or through Data Read muxes. Similar to Data Write muxes, Data Read muxes are commonly used to accommodate multiple-part configurations with a single design. An example of this is shown in Figure 5.13. This schematic of a Data Read mux block is similar to that found in Figure 5.6 for the Data Write mux block. For x16 operation, each output buffer has access to only one Data Read line pair *(DR<n>* and *DR*<n>).* For x8 operation, the eight output buffers each have two pairs of *DR<n>* lines available, doubling the quantity of mbits accessible by each output. Similarly, for x4 operation, the four output buffers have four pairs of *DR<n>* lines available, again doubling the quantity of mbits available for each output. For those configurations with multiple pairs available, address lines control which *DR<n>* pair is connected to an output buffer.

5.1.8 Output Buffer Circuit

The final element in our Read Data path is the output buffer circuit. It consists of an output latch and an output driver circuit. A schematic for an output buffer circuit is shown in Figure 5.14. The output driver on the right side of Figure 5.14 uses three NMOS transistors to drive the output pad to either V_{CCX} or ground. V_{CCX} is the external supply voltage to the DRAM, which may or may not be the same as V_{CC}, depending on whether or not the part is internally regulated. Output drivers using only NMOS transistors are common because they offer better latch-up immunity and ESD protection than CMOS output drivers. Nonetheless, PMOS transistors are still used in CMOS output drivers by several DRAM manufacturers, primarily because they are much easier to drive than full NMOS stages. CMOS outputs are also more prevalent on high-speed synchronous and double-data rate (DDR) designs because they operate at high data rates with less duty-cycle degradation.

In Figure 5.14, two NMOS transistors are placed in series with V_{CCX} to reduce substrate injection currents. Substrate injection currents result from impact ionization, occurring most commonly when high drain-to-source and high gate-to-source voltages exist concurrently. These conditions usually occur when an output driver is firing to V_{CCX}, especially for high-capacitance loads, which slow the output transition. Two transistors in series reduce this effect by lowering the voltages across any single device. The output stage is tristated whenever both signals *PULLUP* and *PULLDOWN* are at ground.

130 Chap. 5 Global Circuitry and Considerations

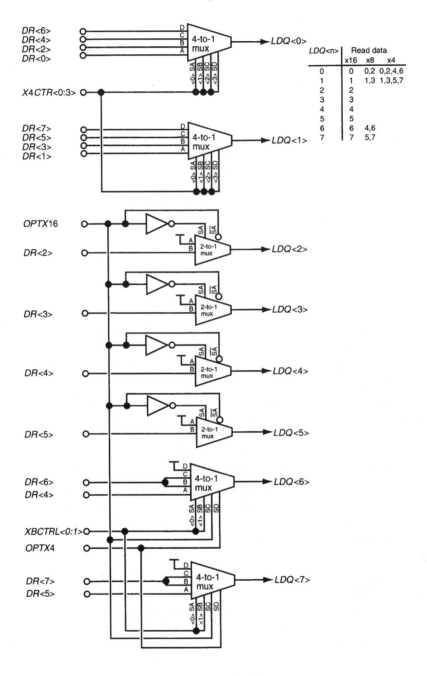

Figure 5.13 Data Read mux.

Sec. 5.1 Data Path Elements 131

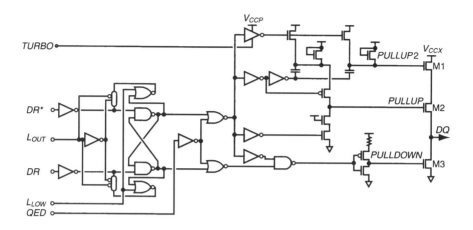

Figure 5.14 Output buffer.

The signal *PULLDOWN* is driven by a simple CMOS inverter, whereas *PULLUP* is driven by a complex circuit that includes voltage charge pumps. The pumps generate a voltage to drive *PULLUP* higher than one V_{TH} above V_{CCX}. This is necessary to ensure that the series output transistors drive the pad to V_{CCX}. The output driver is enabled by the signal labeled *QED*. Once enabled, it remains tristated until either *DQ* or *DQ** fires LOW. If *DR* fires LOW, *PULLDOWN* fires HIGH, driving the pad to ground through M3. If *DR** fires LOW, *PULLUP* fires HIGH, driving the pad to V_{CCX} through M1 and M2.

The output latch circuit shown in Figure 5.14 controls the output driver operation. As the name implies, it contains a latch to hold the output data state. The latch frees the DCSA or HFF and other circuits upstream to get subsequent data for the output. It is capable of storing not only one and zero states, but also a high-impedance state *(tristate)*. It offers transparent operation to allow data to quickly propagate to the output driver. The input to this latch is connected to the *DR*<n> signals coming from either the *DCSA*s or Data Read muxes. Output latch circuits appear in a variety of forms, each serving the needs of a specific application or architecture. The data path may contain additional latches or circuits in support of special modes such as burst operation.

5.1.9 Test Modes

Address compression and data compression are two special test modes that are usually supported by the data path design. Test modes are included in a DRAM design to extend test capabilities or speed component testing or to subject a part to conditions that are not seen during normal operation. Compression test modes yield shorter test times by allowing data from multiple array locations to be tested and compressed on-chip, thereby reducing the effective memory size by a factor of 128 or more in some cases. Address compression, usually on the order of 4x to 32x, is accomplished by internally treating certain address bits as "don't care" addresses.

The data from all of the "don't care" address locations, which correspond to specific data input/output pads *(DQ* pins*)*, are compared using special match circuits. Match circuits are usually realized with NAND and NOR logic gates or through P&E-type drivers on the differential *DR*<n> buses. The match circuits determine if the data from each address location is the same, reporting the result on the respective *DQ* pin as a match or a fail. The data path must be designed to support the desired level of address compression. This may necessitate more DCSA circuits, logic, and pathways than are necessary for normal operation.

The second form of test compression is data compression: combining data at the output drivers. Data compression usually reduces the number of *DQ* pins to four. This compression reduces the number of tester pins required for each part and increases the throughput by allowing additional parts to be tested in parallel. In this way, x16 parts accommodate 4x data compression, and x8 parts accommodate 2x data compression. The cost of any additional circuitry to implement address and data compression must be balanced against the benefits derived from test time reduction. It is also important that operation in test mode correlate 100% with operation in non-test mode. Correlation is often difficult to achieve, however, because additional circuitry must be activated during compression, modifying noise and power characteristics on the die.

5.2 ADDRESS PATH ELEMENTS

DRAMs have used multiplexed addresses since the 4kb generation. Multiplexing is possible because DRAM operation is sequential: column operations follow row operations. Obviously, the column address is not needed until the sense amplifiers have latched, which cannot occur until some time after the wordline has fired. DRAMs operate at higher current levels with multiplexed addressing because an entire page (row address) must be opened with each row access. This disadvantage is overcome by the lower packaging cost associated with multiplexed addresses. In addition, owing to the presence of the *column address strobe ($\overline{CAS}$)*, column operation is independent of row operation, enabling a page to remain open for multiple, high-speed column accesses. This page mode type of operation improves system performance because column access time is much shorter than row access time. Page mode operation appears in more advanced forms, such as EDO and synchronous burst mode, providing even better system performance through a reduction in effective column access time.

The address path for a DRAM can be broken into two parts: the row address path and the column address path. The design of each path is dictated by a unique set of requirements. The address path, unlike the data path, is unidirectional, with address information flowing only into the DRAM. The address path must achieve a high level of performance with minimal power and die area just like any other aspect of DRAM design. Both paths are designed to minimize propagation delay and maximize DRAM performance. In this chapter, we discuss various elements of the row and column address paths.

5.2.1 Row Address Path

The row address path encompasses all of the circuits from the address input pad to the wordline driver. These circuits generally include the address buffer, $\overline{CAS}$ before $\overline{RAS}$ (CBR) counter, predecode logic, array buffers, redundancy logic, phase drivers, and row decoder blocks. Row decoder blocks are discussed in Section 2.3, while redundancy is addressed in Section 4.2. We will now focus on the remaining elements of the row address path, namely, the row address buffer, *CBR* counter, predecode logic, array buffers, and phase drivers.

5.2.2 Row Address Buffer

The row address buffer, as shown schematically in Figure 5.15, consists of a standard input buffer and the additional circuits necessary to implement the functions required for the row address path. The address input buffer *(inpBuf)* is the same as that used for the data input buffer (see Figure 5.1) and must meet the same criteria for V_{IH} and V_{IL} as the data input buffer. The row address buffer includes an inverter latch circuit, as shown in Figure 5.15. This latch, consisting of two inverters and any number of input muxes (two in this case), latches the row address after $\overline{RAS}$ falls.

The input buffer drives through a mux, which is controlled by a signal called *row address latch (RAL)*. Whenever *RAL* is LOW, the mux is enabled. The feedback inverter has low drive capability, allowing the latch to be overwritten by either the address input buffer or the *CBR* counter, depending on which mux is enabled. The latch circuit drives into a pair of NAND gates, forcing both *RA*<n> and *RA**<n> to logic LOW states whenever the row address buffer is disabled because *RAEN* is LOW.

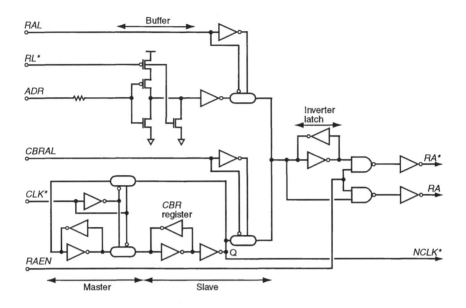

Figure 5.15 Row address buffer.

5.2.3 CBR Counter

As illustrated in Figure 5.15, the *CBR* ($\overline{CAS}$ before $\overline{RAS}$) counter consists of a single inverter and a pair of inverter latches coupled through a pair of complementary muxes to form a one-bit counter. For every HIGH-to-LOW transition of *CLK**, the register output at Q toggles. All of the *CBR* counters from each row address buffer are cascaded together to form a *CBR* ripple counter. The Q output of one stage feeds the *CLK** input of a subsequent stage. The first register in the counter is clocked whenever $\overline{RAS}$ falls while in a *CBR* Refresh mode. By cycling through all possible row address combinations in a minimum of clocks, the *CBR* ripple counter provides a simple means of internally generating Refresh addresses. The *CBR* counter drives through a mux into the inverter latch of the row address buffer. This mux is enabled whenever *CBR address latch (CBRAL)* is LOW. Note that the signals *RAL* and *CBRAL* are mutually exclusive in that they cannot be LOW at the same time. For each and every DRAM design, the row address buffer and *CBR* counter designs take on various forms. Logic may be inverted, counters may be more or less complex, and muxes may be replaced with static gates. Whatever the differences, however, the function of the input buffer and its *CBR* counter remains essentially the same.

Sec. 5.2 Address Path Elements   135

5.2.4 Predecode Logic

As discussed in earlier sections of this book, using predecoded addressing internal to a DRAM has many advantages: lower power, higher efficiency, and a simplified layout. An example of predecode circuits for the row address path is shown in Figure 5.16. This schematic consists of seven sets of predecoders. The first set is used for even and odd row selection and consists only of cascaded inverter stages. $RA<0>$ is not combined with other addresses due to the nature of row decoding. In our example, we assume that odd and even will be combined with the predecoded row addresses at a later point in the design, such as at the array interfaces. The next set of predecoders is for addresses $RA<1>$ and $RA<2>$, which together form $RA\,12<0:3>$. Predecoding is accomplished through a set of two-input NAND gates and inverter buffers, as shown. The remaining addresses are identically decoded except for $RA<12>$. As illustrated at the bottom of the schematic, $RA<12>$ and $RA^*<12>$ are taken through a NOR gate circuit. This circuit forces both $RA<12>$ lines to a HIGH state as they enter the decoder, whenever the DRAM is configured for 4k Refresh. This process essentially makes $RA<12>$ a "don't care" address in the predecode circuit, forcing twice as many wordlines to fire at a time.

5.2.5 Refresh Rate

Normally, when Refresh rates change for a DRAM, a higher order address is treated as a "don't care" address, thereby decreasing the row address space but increasing the column address space. For example, a 16Mb DRAM bonded as a 4Mb x4 part could be configured in several Refresh rates: 1k, 2k, and 4k. Table 5.1 shows how row and column addressing is related to these Refresh rates for the 16Mb example. In this example, the 2k Refresh rate would be more popular because it has an equal number of row and column addresses or *square* addressing.

Refresh rate is also determined by backward compatibility, especially in personal computer designs. Because a 4Mb DRAM has less memory space than a 16Mb DRAM, the 4Mb DRAM should naturally have fewer address pins. To sell 16Mb DRAMs into personal computers that are designed for 4Mb DRAMs, the 16Mb part must be configured with no more address pins than the 4Mb part. If the 4Mb part has eleven address pins, then the 16Mb part should have only eleven address pins, hence 2k Refresh. To trim cost, most PC designs keep the number of DRAM address pins to a minimum. Although this practice holds cost down, it also limits expandability and makes conversion to newer DRAM generations more complicated owing to resultant backward compatibility issues.

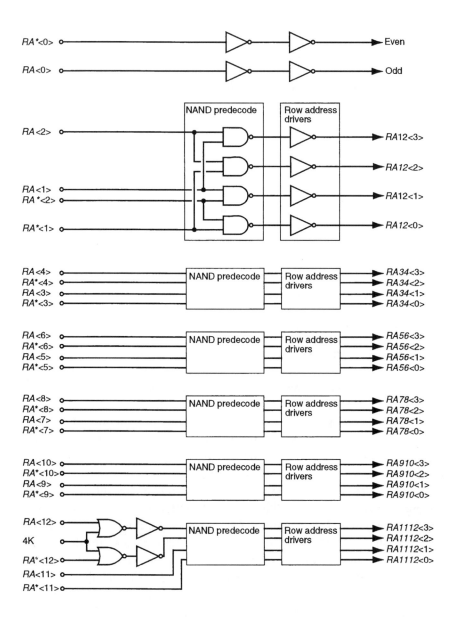

Figure 5.16 Row address predecode circuits.

Sec. 5.2 Address Path Elements 137

Table 5.1 Refresh rate versus row and column addresses.

Refresh Rate	Rows	Columns	Row Addresses	Column Addresses
4K	4,096	1,024	12	10
2K	2,048	2,048	11	11
1K	1,024	4,096	10	12

5.2.6 Array Buffers

The next elements to be discussed in the row address path are array buffers and phase drivers. The array buffers drive the predecoded address signals into the row decoder blocks. In general, the buffers are no more than cascaded inverters, but in some cases they include static logic gates or level translators, depending on row decoder requirements. Additional logic gates could be included for combining the addresses with enabling signals from the control logic or for making odd/even row selection by combining the addresses with the odd and even address signals. Regardless, the resulting signals ultimately drive the decode trees, making speed an important issue. Buffer size and routing resistance, therefore, become important design parameters in high-speed designs because the wordline cannot be fired until the address tree is decoded and ready for the *PHASE* signal to fire.

5.2.7 Phase Drivers

As the discussion concerning wordline drivers and tree decoders in Section 2.3 showed, the signal that actually fires the wordline is called *PHASE*. Although the signal name may vary from company to company, the purpose of the signal does not. Essentially, this signal is the final address term to arrive at the wordline driver. Its timing is carefully determined by the control logic. *PHASE* cannot fire until the row addresses are set up in the decode tree. Normally, the timing of *PHASE* also includes enough time for the row redundancy circuits to evaluate the current address. If a redundancy match is found, the normal row cannot be fired. In most DRAM designs, this means that the normally decoded *PHASE* signal will not fire but will instead be replaced by some form of redundant *PHASE* signal.

A typical phase decoder/driver is shown in Figure 5.17. Again, like so many other DRAM circuits, it is composed of standard static logic gates. A level translator would be included in the design if the wordline driver required a *PHASE* signal that could drive to a boosted voltage supply. This translator could be included with the phase decoder logic or placed in array gaps to locally drive selected row decoder blocks. Local phase translators are common on double-metal designs with local row decoder blocks.

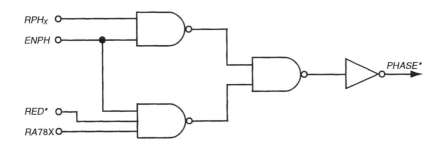

Figure 5.17 Phase decoder/driver.

5.2.8 Column Address Path

With our examination of row address path elements complete, we can turn our attention to the column address path. The column address path consists of input buffers, address transition detection circuits, predecode logic, redundancy, and column decode circuits. Redundancy and column decode circuits are addressed in Section 4.2 and Section 4.1, respectively.

A column address buffer schematic, as shown in Figure 5.18, consists of an input buffer, a latch, and address transition circuits. The input buffer shown in this figure is again identical to that described in Section 5.1 for the data path, so further description is unnecessary. The column address input buffer, however, is disabled by the signal *power column (PCOL*)* whenever $\overline{RAS}$ is HIGH and the part is inactive. Normally, the column address buffers are enabled by *PCOL** shortly after $\overline{RAS}$ goes LOW. The input buffer feeds into a NAND latch circuit, which traps the column address whenever the *column address latch (CAL*)* fires LOW. At the start of a column cycle, *CAL** is HIGH, making the latch transparent. This transparency permits the column address to propagate through to the predecode circuits. The latch output also feeds into an *address transition detection (ATD)* circuit that is shown on the right side of Figure 5.18.

Sec. 5.2 Address Path Elements 139

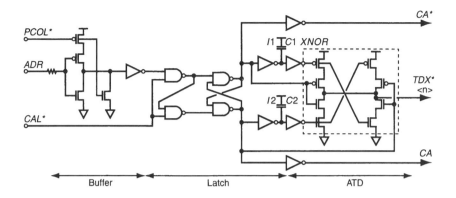

Figure 5.18 Column address buffer.

5.2.9 Address Transition Detection

The address transition detection *(ATD)* circuit is extremely important to page mode operation in a DRAM. An *ATD* circuit detects any transition that occurs on a respective address pin. Because it follows the NAND latch circuit, the *CAL** control signal must be HIGH. This signal makes the latch transparent and thus the *ATD* functional. The *ATD* circuit in Figure 5.18 has symmetrical operation: it can sense either a rising or a falling edge. The circuit uses a two-input XNOR gate to generate a pulse of prescribed duration whenever a transition occurs. Pulse duration is dictated by simple delay elements: I1 and C1 or I2 and C2. Both delayed and undelayed signals from the latch are fed into the XNOR gate. The *ATD* output signals from all of the column addresses, labeled *TDX**<n>, are routed to the equilibration driver circuit shown in Figure 5.19. This circuit generates a set of equilibration signals for the DRAM. The first of these signals is *equilibrate I/O (EQIO*)*, which, as the name implies, is used in arrays to force equilibration of the *I/O* lines. As we learned in Section 5.1.4, the I/O lines need to be equilibrated to $V_{CC} - V_{TH}$ prior to a new column being selected by the *CSEL*<n> lines. *EQIO** is the signal used to accomplish this equilibration.

The second signal generated by the equilibration driver is called *equilibrate sense amp (EQSA)*. This signal is generated from address transitions occurring on all of the column addresses, including the *least significant* addresses. The least significant column addresses are not decoded into the *column select lines (CSEL)*. Rather, they are used to select which set of *I/O* lines is connected to the output buffers. As shown in the schematic, *EQSA* is activated regardless of which address is changed because the *DCSA*s must be equilibrated prior to sensing any new data. *EQIO*, on the other hand, is not affected by the least significant addresses because the *I/O* lines do not need equilibrating unless the *CSEL* lines are changed. The equilibration driver circuit, as shown in Figure 5.19, uses a balanced NAND gate to combine the pulses from each *ATD* circuit. Balanced logic helps ensure that the narrow *ATD* pulses are not distorted as they progress through the circuit.

The column addresses are fed into predecode circuits, which are very similar to the row address predecoders. One major difference, however, is that the column addresses are not allowed to propagate through the part until the wordline has fired. For this reason, the signal *Enable column (ECOL)* is gated into the predecode logic as shown in Figure 5.20. ECOL disables the predecoders whenever it is LOW, forcing the outputs all HIGH in our example. Again, the predecode circuits are implemented with simple static logic gates. The address signals emanating from the predecode circuits are buffered and distributed throughout the die to feed the column decoder logic blocks. The column decoder elements are described in Section 4.1.

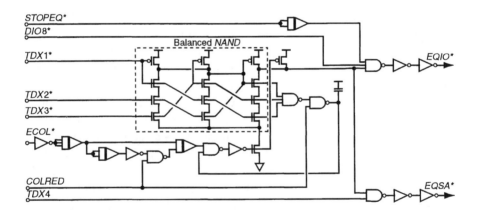

Figure 5.19 Equilibration driver.

Sec. 5.2 Address Path Elements

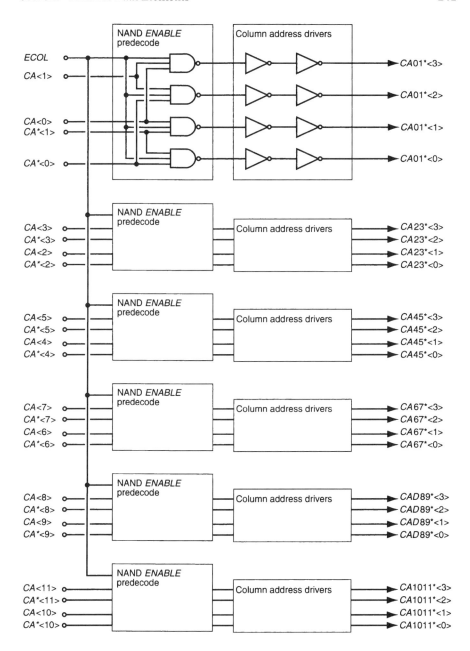

Figure 5.20 Column predecode logic.

5.3 SYNCHRONIZATION IN DRAMS[1]

In a typical SDRAM, the relationship when reading data out of the DRAM between the *CLK* input and the valid data out time or *access time* t_{AC} can vary widely with process shifts, temperature, and operating clock frequency. Figure 5.21 shows a typical relationship between the SDRAM, *CLK* input, and a *DQ* output. The parameter specification t_{AC} is generally specified as being less than some value, for example, t_{AC} < 5ns.

As clock frequencies increase, it is desirable to have less uncertainty in the availability of valid data on the output of the DRAM. Towards this goal, double data rate (DDR) SDRAMs use a *delay-locked loop* (DLL) [3] to drive t_{AC} to zero. (A typical specification for t_{AC} in a DDR SDRAM is ± 0.75ns.) Figure 5.22 shows the block diagram for a DLL used in a DDR SDRAM. Note that the data *I/O* in DDR is clocked on both the rising and falling edges of the input *CLK* (actually on the output of the delay line) [4]. Also, note that the input *CLK* should be synchronized with the *DQ strobe (DQS)* clock output.

To optimize and stabilize the clock-access and output-hold times in an SDRAM, an internal *register-controlled delay-locked loop* (RDLL) has been used [4][5][6]. The RDLL adjusts the time difference between the output (*DQ*s) and input (*CLK*) clock signals in SDRAM until they are aligned. Because the RDLL is an all-digital design, it provides robust operation over all process corners. Another solution to the timing constraints found in SDRAM was given by using the *synchronous mirror delay* (SMD) in [7]. Compared to RDLL, the SMD does not lock as tightly, but the time to acquire lock between the input and output clocks is only two clock cycles.

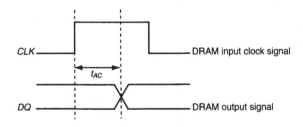

Figure 5.21 SDRAM *CLK* input and *DQ* output.

---

[1] This material is taken directly from [4].

Sec. 5.3 Synchronization in DRAMs 143

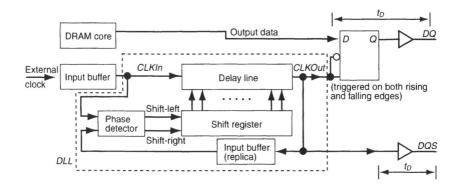

Figure 5.22 Block diagram for DDR SDRAM DLL.

As DRAM clock speeds continue to increase, the skew becomes the dominating concern, outweighing the RDLL disadvantage of longer time to acquire lock.

This section describes an RSDLL (register-controlled symmetrical DLL), which meets the requirements of DDR SDRAM. (Read/Write accesses occur on both rising and falling edges of the clock.) Here, *symmetrical* means that the delay line used in the DLL has the same delay whether a HIGH-to-LOW or a LOW-to-HIGH logic signal is propagating along the line. The data output timing diagram of a DDR SDRAM is shown in Figure 5.23. The RSDLL increases the valid output data window and diminishes the undefined t_{DSDQ} by synchronizing both the rising and falling edges of the *DQS* signal with the output data *DQ*.

Figure 5.22 shows the block diagram of the RSDLL. The replica input buffer dummy delay in the feedback path is used to match the delay of the input clock buffer. The *phase detector (PD)* compares the relative timing of the edges of the input clock signal and the feedback clock signal, which comes through the delay line and is controlled by the shift register. The outputs of the *PD*, shift-right and shift-left, control the shift register. In the simplest case, one bit of the shift register is HIGH. This single bit selects a point of entry for *CLKIn* the symmetrical delay line. (More on this later.) When the rising edge of the input clock is within the rising edges of the output clock and one unit delay of the output clock, both outputs of the *PD*, shift-right and shift-left, go LOW and the loop is locked.

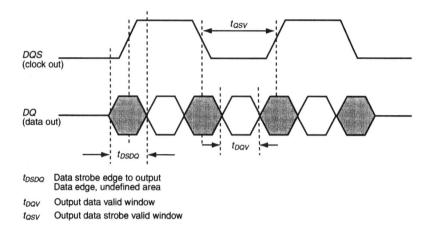

t_{DSDQ} Data strobe edge to output
Data edge, undefined area
t_{DQV} Output data valid window
t_{QSV} Output data strobe valid window

Figure 5.23 Data timing chart for DDR DRAM.

5.3.1 The Phase Detector

The basic operation of the *phase detector (PD)* is shown in Figure 5.24. The resolution of this RSDLL is determined by the size of the unit delay used in the delay line. The locking range is determined by the number of delay stages used in the symmetrical delay line. Because the DLL circuit inserts an optimum delay time between *CLKIn* and *CLKOut*, making the output clock change simultaneously with the next rising edge of the input clock, the minimum operating frequency to which the RSDLL can lock is the reciprocal of the product of the number of stages in the symmetrical delay line with the delay per stage. Adding more delay stages increases the locking range of the RSDLL at the cost of increased layout area.

5.3.2 The Basic Delay Element

Rather than using an AND gate as the unit-delay stage (NAND + inverter), as in [5], a NAND-only-based delay element can be used. The implementation of a three-stage delay line is shown in Figure 5.25. The problem when using a NAND + inverter as the basic delay element is that the propagation delay through the unit delay resulting from a HIGH-to-LOW transition is not equal to the delay of a LOW-to-HIGH transition ($t_{PHL} \neq t_{PLH}$). Furthermore, the delay varies from one run to another. If the skew between t_{PHL} and t_{PLH} is 50 ps, for example, the total skew of the falling edges through ten stages will be 0.5 ns. Because of this skew, the NAND + inverter delay element cannot be used in a DDR DRAM. In our modified symmetrical delay element, another NAND gate is used instead of an

Sec. 5.3 Synchronization in DRAMs

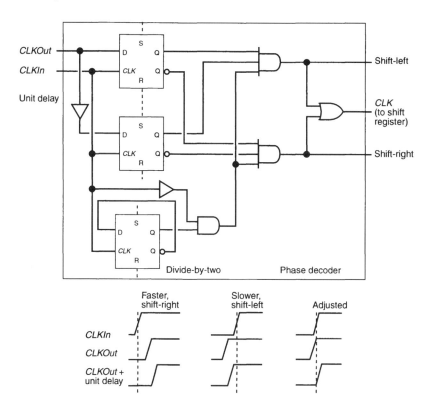

Figure 5.24 Phase detector used in RSDLL.

inverter (two NAND gates per delay stage). This scheme guarantees that t_{PHL} = t_{PLH} independent of process variations. While one NAND switches from HIGH to LOW, the other switches from LOW to HIGH. An added benefit of the two-NAND delay element is that two point-of-entry control signals are now available. The shift register uses both to solve the possible problem caused by the POWERUP ambiguity in the shift register.

5.3.3 Control of the Shift Register

As shown in Figures 5.25 and 5.26, the input clock is a common input to every delay stage. The shift register selects a different *tap* of the delay line (the point of entry for the input clock signal into the symmetrical delay line). The complementary outputs of each register cell select the different tap: Q is connected directly to the input A of a delay element, and Q^* is connected to the previous stage of input B.

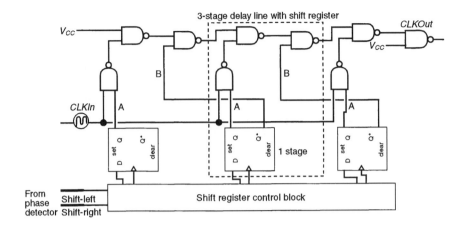

Figure 5.25 Symmetrical delay element used in RSDLL.

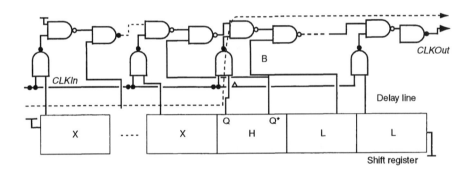

Figure 5.26 Delay line and shift register for RSDLL.

From right to left, the first LOW-to-HIGH transition in the shift register sets the point of entry into the delay line. The input clock passes through the tap with a HIGH logic state in the corresponding position of the shift register. Because the Q^* of this tap is equal to a LOW, it disables the previous stages; therefore, the previous states of the shift register do not matter (shown as "don't care," X, in Figure 5.25). This control mechanism guarantees that only one path is selected. This scheme also eliminates POWERUP concerns because the selected tap is simply the first, from the right, LOW-to-HIGH transition in the register.

5.3.4 Phase Detector Operation

To stabilize the movement in the shift register, after making a decision, the phase detector waits at least two clock cycles before making another

Sec. 5.3 Synchronization in DRAMs 147

decision (Figure 5.24). A divide-by-two is included in the phase detector so that every other decision, resulting from comparing the rising edges of the external clock and the feedback clock, is used. This provides enough time for the shift register to operate and the output waveform to stabilize before another decision by the *PD* is implemented. The unwanted side effect of this delay is an increase in lock time. The shift register is clocked by combining the shift-left and -right signals. The power consumption decreases when there are no shift-left or -right signals and the loop is locked.

Another concern with the phase-detector design is the design of the flip-flops (FFs). To minimize the static phase error, very fast FFs should be used, ideally with zero setup time.

Also, the metastability of the flip-flops becomes a concern as the loop locks. This, together with possible noise contributions and the need to wait, as discussed above, before implementing a shift-right or -left, may make it more desirable to add more filtering in the phase detector. Some possibilities include increasing the divider ratio of the phase detector or using a shift register in the phase detector to determine when a number of—say, four—shift-rights or -lefts have occurred. For the design in Figure 5.26, a divide-by-two was used in the phase detector due to lock-time requirements.

5.3.5 Experimental Results

The RSDLL of Figure 5.22 was fabricated in a 0.21 µm, four-poly, double-metal CMOS technology (a DRAM process). A 48-stage delay line with an operation frequency of 125–250 MHz was used. The maximum operating frequency was limited by delays external to the DLL, such as the input buffer and interconnect. There was no noticeable static phase error on either the rising or falling edges. Figure 5.27 shows the resulting rms jitter versus input frequency. One sigma of jitter over the 125–250 MHz frequency range was below 100 ps. The measured delay per stage versus V_{CC} and temperature is shown in Figure 5.28. Note that the 150 ps typical delay of a unit-delay element was very close to the rise and fall times on-chip of the clock signals and represents a practical minimum resolution of a DLL for use in a DDR DRAM fabricated in a 0.21 µm process.

The power consumption (the current draw of the DLL when V_{CC} = 2.8 V) of the prototype RSDLL is illustrated in Figure 5.29. It was found that the power consumption was determined mainly by the dynamic power dissipation of the symmetrical delay line. The NAND delays in this test chip were implemented with 10/0.21 µm NMOS and 20/0.21 µm PMOS. By reducing the widths of both the NMOS and PMOS transistors, the power dissipation is greatly reduced without a speed or resolution penalty (with the added benefit of reduced layout size).

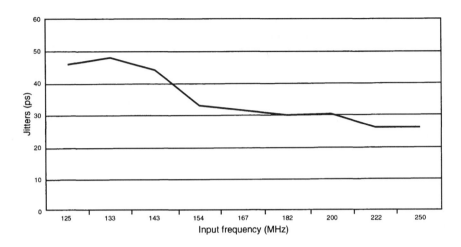

Figure 5.27 Measured rms jitter versus input frequency.

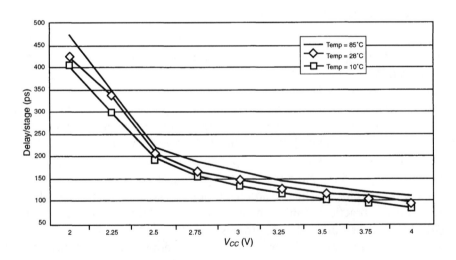

Figure 5.28 Measured delay per stage versus V_{CC} and temperature.

Sec. 5.3 Synchronization in DRAMs

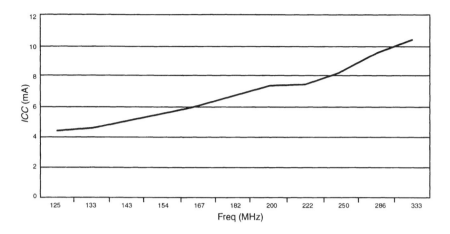

Figure 5.29 Measured *ICC* (DLL current consumption) versus input frequency.

5.3.6 Discussion

In this section we have presented one possibility for the design of a delay-locked loop. While there are others, this design is simple, manufacturable, and scalable.

In many situations the resolution of the phase detector must be decreased. A useful circuit to determine which one of two signals occurs earlier in time is shown in Figure 5.30. This circuit is called an arbiter. If *S*1 occurs slightly before *S*2 then the output *SO*1 will go HIGH, while the output *SO*2 stays LOW. If *S*2 occurs before *S*1, then the output *SO*2 goes HIGH and *SO*1 remains LOW. The fact that the inverters on the outputs are powered from the SR latch (the cross-coupled NAND gates) ensures that *SO*1 and *SO*2 cannot be HIGH at the same time. When designed and laid out correctly, this circuit is capable of discriminating tens of picoseconds of difference between the rising edges of the two input signals.

The arbiter alone cannot be capable of controlling the shift register. A simple logic block to generate shift-right and shift-left signals is shown in Figure 5.31. The rising edge of *SO*1 or *SO*2 is used to clock two D-latches so that the shift-right and shift-left signals may be held HIGH for more than one clock cycle. Figure 5.31 uses a divide-by-two to hold the shift signals valid for two clock cycles. This is important because the output of the arbiter can have glitches coming from the different times when the inputs go back LOW. Note that using an arbiter-based phase detector alone can result in an alternating sequence of shift-right, shift-left. We eliminated this problem in the phase-detector of Figure 5.24 by introducing the dead zone so that a minimum delay spacing of the clocks would result in no shifting.

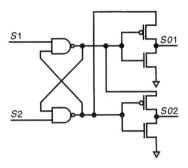

Figure 5.30 Two-way arbiter as a phase detector.

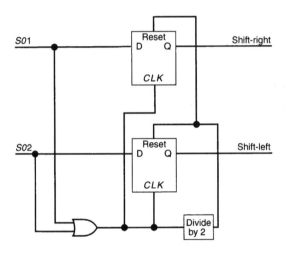

Figure 5.31 Circuit for generating shift register control.

In some situations, the fundamental delay time of an element in the delay line needs reduction. Using the NAND-based delay shown in Figure 5.26, we are limited to delay times much longer than a single inverter delay. However, using a single inverter delay results in an inversion in our basic cell. Figure 5.32 shows that using a double inverter-based delay element can solve this problem. By crisscrossing the single delay element inputs the cell appears to be non-inverting. This scheme results in the least delay because the delay between the cell's input and output is only the delay of a single inverter. The problems with using this type of cell over the NAND-based delay are inserting the feedback clock and the ability of the shift register to control the delay of the resulting delay line.

Sec. 5.3 Synchronization in DRAMs 151

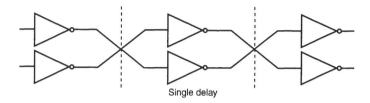

Figure 5.32 A double inverter used as a delay element.

Figure 5.33 shows how inserting transmission gates (TGs) that are controlled by the shift register allows the insertion point to vary along the line. When C is HIGH, the feedback clock is inserted into the output of the delay stage. The inverters in the stage are isolated from the feedback clock by an additional set of TGs. We might think, at first glance, that adding the TGs in Figure 5.33 would increase the delay significantly; however, there is only a single set of TGs in series with the feedback before the signal enters the line. The other TGs can be implemented as part of the inverter to minimize their impact on the overall cell delay. Figure 5.34 shows a possible inverter implementation.

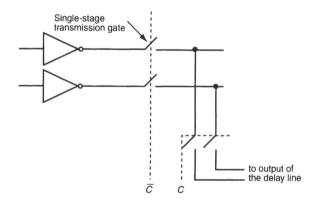

Figure 5.33 Transmission gates added to delay line.

Finally, in many situations other phases of an input clock need generating. This is especially useful in minimizing the requirements on the setup and hold times in a synchronous system. Figure 5.35 shows a method of segmenting the delays in a DLL delay line. A single control register can be used for all four delay segments. The challenge becomes reducing the amount of delay in a single delay element. A shift in this segmented configuration results in a change in the overall delay of four delays rather than a single delay. Note that with a little thought, and realizing that the delay elements of Figure 5.32 can be used in inverting or non-inverting configurations, the delay line of Figure 5.35 can be implemented with only two segments and still provide taps of 90°, 180°, 270°, and 360°.

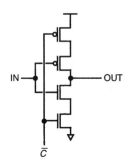

Figure 5.34 Inverter implementation.

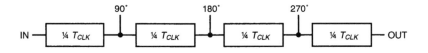

Figure 5.35 Segmenting delays for additional clocking taps.

REFERENCES

[1] H. Ikeda and H. Inukai, "High-Speed DRAM Architecture Development," *IEEE Journal of Solid-State Circuits*, vol. 34, no. 5, pp. 685–692, May 1999.

[2] M. Bazes, "Two Novel Full Complementary Self-Biased CMOS Differential Amplifiers," *IEEE Journal of Solid-State Circuits*, vol. 26, no. 2, pp. 165–168, February 1991.

[3] R. J. Baker, H. W. Li, and D. E. Boyce, *CMOS: Circuit Design, Layout, and Simulation*, Piscataway, NJ: IEEE Press, 1998.

[4] F. Lin, J. Miller, A. Schoenfeld, M. Ma, and R. J. Baker, "A Register-Controlled Symmetrical DLL for Double-Data-Rate DRAM," *IEEE Journal of Solid-State Circuits*, vol. 34, no. 4, 1999.

[5] A. Hatakeyama, H. Mochizuki, T. Aikawa, M. Takita, Y. Ishii, H. Tsuboi, S. Y. Fujioka, S. Yamaguchi, M. Koga, Y. Serizawa, K. Nishimura, K. Kawabata, Y. Okajima, M. Kawano, H. Kojima, K. Mizutani, T. Anezaki,

M. Hasegawa, and M. Taguchi, "A 256-Mb SDRAM Using a Register-Controlled Digital DLL," *IEEE Journal of Solid-State Circuits,* vol. 32, pp. 1728–1732, November 1997.

[6] S. Eto, M. Matsumiya, M. Takita, Y. Ishii, T. Nakamura, K. Kawabata, H. Kano, A. Kitamoto, T. Ikeda, T. Koga, M. Higashiro, Y. Serizawa, K. Itabashi, O. Tsuboi, Y. Yokoyama, and M. Taguchi, "A 1Gb SDRAM with Ground Level Precharged Bitline and Non-Boosted 2.1V Word Line," *ISSCC Digest of Technical Papers,* pp. 82–83, February 1998.

[7] T. Saeki, Y. Nakaoka, M. Fujita, A. Tanaka, K. Nagata, K. Sakakibara, T. Matano, Y. Hoshino, K. Miyano, S. Isa, S. Nakazawa, E. Kakehashi, J. M. Drynan, M. Komuro, T. Fukase, H. Iwasaki, M. Takenaka, J. Sekine, M. Igeta, N. Nakanishi, T. Itani, K. Yoshida, H. Yoshino, S. Hashimoto, T. Yoshii, M. Ichinose, T. Imura, M. Uziie, S. Kikuchi, K. Koyama, Y. Fukuzo, and T. Okuda, "A 2.5-ns Clock Access 250-MHz, 256-Mb SDRAM with Synchronous Mirror Delay," *IEEE Journal of Solid-State Circuits,* vol. 31, pp. 1656–1665, November 1996.

Chapter 6

Voltage Converters

In this chapter, we discuss the circuitry for generating the on-chip voltages that lie outside the supply range. In particular, we look at the wordline pump voltage and the substrate pumps. We also discuss voltage regulators that generate the internal power supply voltages.

6.1 INTERNAL VOLTAGE REGULATORS

6.1.1 Voltage Converters

DRAMs depend on a variety of internally generated voltages to operate and to optimize their performance. These voltages generally include the boosted wordline voltage V_{CCP}, the internally regulated supply voltage V_{CC}, the $V_{CC}/2$ cellplate and digitline bias voltage $DVC2$, and the pumped substrate voltage V_{BB}. Each of these voltages is regulated with a different kind of voltage generator. A linear voltage converter generates V_{CC}, while a modified CMOS inverter creates $DVC2$. Generating the boosted supply voltages V_{CCP} and V_{BB} requires sophisticated circuits that employ charge on voltage pumps (a.k.a., charge pumps). In this chapter, we discuss each of these circuits and how they are used in modern DRAM designs to generate the required supply voltages.

Most modern DRAM designs rely on some form of internal voltage regulation to convert the external supply voltage V_{CCX} into an internal supply voltage V_{CC}. We say most, not all, because the need for internal regulation is dictated by the external voltage range and the process in which the DRAM is based. The process determines gate oxide thickness, field device characteristics, and diffused junction properties. Each of these properties, in turn, affects breakdown voltages and leakage parameters, which limit the maximum operating voltage that the process can reliably tolerate. For example, a 16Mb DRAM built in a 0.35 µm CMOS process with a 120Å thick gate oxide can operate reliably with an internal supply voltage not exceeding 3.6 V. If this design had to operate in a 5 V system, an internal voltage regulator would be needed to convert the external 5 V supply to an internal 3.3 V supply. For the same design operating in a 3.3 V system, an internal voltage regulator would not be required. Although the actual operating voltage is determined by process considerations and reliability studies, the internal supply voltage is generally proportional to the minimum feature size. Table 6.1 summarizes this relationship.

Table 6.1 DRAM process versus supply voltage.

Process	V_{CC} Internal
0.45 µm	4 V
0.35 µm	3.3 V
0.25 µm	2.5 V
0.20 µm	2 V

All DRAM voltage regulators are built from the same basic elements: a voltage reference, one or more output power stages, and some form of control circuit. How each of these elements is realized and combined into the overall design is the product of process and design limitations and the design engineer's preferences. In the paragraphs that follow, we discuss each element, overall design objectives, and one or more circuit implementations.

6.1.2 Voltage References

The sole purpose of a voltage reference circuit is to establish a nominal operating point for the voltage regulator circuit. However, this nominal voltage is not a constant; rather, it is a function of the external voltage V_{CCX} and temperature. Figure 6.1 is a graph of V_{CC} versus V_{CCX} for a typical DRAM voltage regulator. This figure shows three regions of operation. The first region occurs during a POWERUP or POWERDOWN cycle, in which V_{CCX} is below the desired V_{CC} operating voltage range. In this region, V_{CC} is set equal to V_{CCX}, providing the maximum operating voltage allowable in

Sec. 6.1 Internal Voltage Regulators 157

the part. A maximum voltage is desirable in this region to extend the DRAM's operating range and ensure data retention during low-voltage conditions.

The second region exists whenever V_{CCX} is in the nominal operating range. In this range, V_{CC} flattens out and establishes a relatively constant supply voltage to the DRAM. Various manufacturers strive to make this region absolutely flat, eliminating any dependence on V_{CCX}. We have found, however, that a moderate amount of slope in this range for characterizing performance is advantageous. It is critically important in a manufacturing environment that each DRAM meet the advertised specifications, with some margin for error. A simple way to ensure these margins is to exceed the operating range by a fixed amount during component testing. The voltage slope depicted in Figure 6.1 allows this margin testing to occur by establishing a moderate degree of dependence between V_{CCX} and V_{CC}.

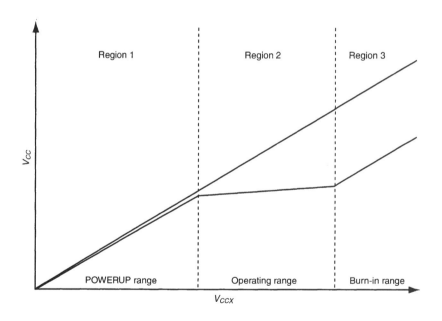

Figure 6.1 Ideal regulator characteristics.

The third region shown in Figure 6.1 is used for component burn-in. During burn-in, both the temperature and voltage are elevated above the normal operating range to stress the DRAMs and weed out infant failures. Again, if there were no V_{CCX} and V_{CC} dependency, the internal voltage could not be elevated. A variety of manufacturers do not use the monotonic curve shown in Figure 6.1. Some designs break the curve as shown in Figure 6.2, producing a step in the voltage characteristics. This step creates a region in which the DRAM cannot be operated. We will focus on the more desirable circuits that produce the curve shown in Figure 6.1.

To design a voltage reference, we need to make some assumptions about the power stages. First, we will assume that they are built as unbuffered, two-stage, CMOS operational amplifiers and that the gain of the first stage is sufficiently large to regulate the output voltage to the desired accuracy. Second, we will assume that they have a closed loop gain of Av. The value of Av influences not only the reference design, but also the operating characteristics of the power stage (to be discussed shortly). For this design example, assume Av = 1.5. The voltage reference circuit shown in Figure 6.3 can realize the desired V_{CC} characteristics shown in Figure 6.4. This circuit uses a simple resistor and a PMOS diode reference stack that is buffered and amplified by an unbuffered CMOS op-amp. The resistor and diode are sized to provide the desired output voltage and temperature characteristics and the minimum bias current. Note that the diode stack is programmed through the series of PMOS switch transistors that are shunting the stack. A fuse element is connected to each PMOS switch gate. Unfortunately, this programmability is necessary to accommodate process variations and design changes.

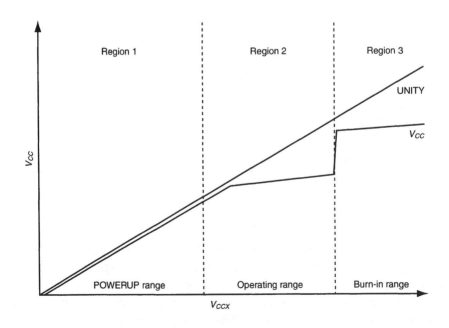

Figure 6.2 Alternative regulator characteristics.

Sec. 6.1 Internal Voltage Regulators

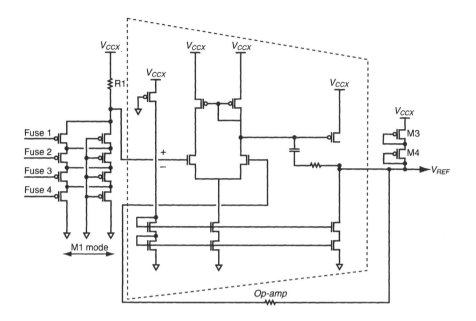

Figure 6.3 Resistor/diode voltage reference.

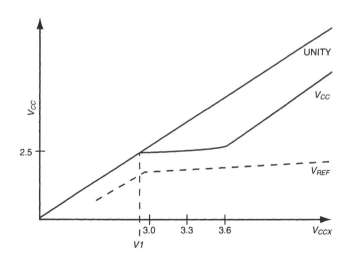

Figure 6.4 Voltage regulator characteristics.

The reference temperature characteristics rely on establishing a proper balance between V_{TH}, mobility, and resistance variations with temperature. An ideal temperature coefficient for this circuit would be positive such that the voltage rises with temperature, somewhat compensating for the CMOS gate delay speed loss associated with increasing temperature.

Two PMOS diodes (M3–M4) connected in series with the op-amp output terminal provide the necessary burn-in characteristics in Region 3. In normal operation, the diodes are OFF. As V_{CCX} is increased into the burn-in range, V_{CCX} will eventually exceed V_{REF} by two diode drops, turning ON the PMOS diodes and clamping V_{REF} to V_{TH} below V_{CCX}. The clamping action will establish the desired burn-in characteristics, keeping the regulator monotonic in nature.

In Region 2, voltage slope over the operating range is determined by the resistance ratio of the PMOS reference diode M1 and the bias resistor R1. Slope reduction is accomplished by either increasing the effective PMOS diode resistance or replacing the bias resistor with a more elaborate current source as shown in Figure 6.5. This current source is based on a V_{TH} referenced source to provide a reference current that is only slightly dependent on V_{CCX} [1]. A slight dependence is still necessary to generate the desired voltage slope.

The voltage reference does not actually generate Region 1 characteristics for the voltage regulator. Rather, the reference ensures a monotonic tran-

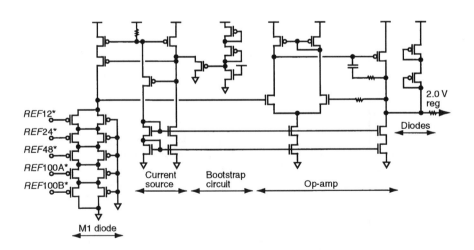

Figure 6.5 Improved voltage reference.

sition from Region 1 to Region 2. To accomplish this task, the reference must approximate the ideal characteristics for Region 1, in which $V_{CC} = V_{CCX}$. The regulator actually implements Region 1 by shorting the V_{CC} and V_{CCX} buses together through the PMOS output transistors found in each power stage op-amp. Whenever V_{CCX} is below a predetermined voltage V1, the PMOS gates are driven to ground, actively shorting the buses together. As V_{CCX} exceeds the voltage level V1, the PMOS gates are released and normal regulator operation commences. To ensure proper DRAM operation, this transition needs to be as seamless as possible.

6.1.3 Bandgap Reference

Another type of voltage reference that is popular among DRAM manufacturers is the bandgap reference. The bandgap reference is traditionally built from vertical pnp transistors. A novel bandgap reference circuit is presented in Figure 6.6. As shown, it uses two bipolar transistors with an emitter size ratio of 10:1. Because they are both biased with the same current and owing to the different emitter sizes, a differential voltage will exist between the two transistors. The differential voltage appearing across resistor R1 will be amplified by the op-amp. Resistors R2 and R1 establish the closed loop gain for this amplifier and determine nominal output voltage and bias currents for the transistors [1].

The almost ideal temperature characteristics of a bandgap reference are what make them attractive to regulator designers. Through careful selection of emitter ratios and bias currents, the temperature coefficient can be set to approximately zero. Also, because the reference voltage is determined by the bandgap characteristics of silicon rather than a PMOS V_{TH}, this circuit is less sensitive to process variations than the circuit in Figure 6.3.

Three problems with the bandgap reference shown in Figure 6.6, however, make it much less suitable for DRAM applications. First, the bipolar transistors need moderate current to ensure that they operate beyond the knee on their I-V curves. This bias current is approximately 10–20 µA per transistor, which puts the total bias current for the circuit above 25 µA. The voltage reference shown in Figure 6.3, on the other hand, consumes less than 10 µA of the total bias current. Second, the vertical pnp transistors inject a significant amount of current into the substrate—as high as 50% of the total bias current in some cases. For a pumped substrate DRAM, the resulting charge from this injected current must be removed by the substrate pump, which raises standby current for the part. Third, the voltage slope for a bandgap reference is almost zero because the feedback configuration in Figure 6.6 has

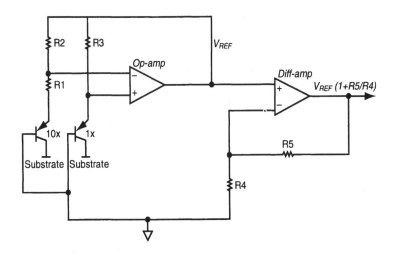

Figure 6.6 Bandgap reference circuit.

no dependence on V_{CCX}. Seemingly ideal, this performance is unacceptable because to perform margin testing on the DRAM a finite voltage slope is needed.

6.1.4 The Power Stage

Although static voltage characteristics of the DRAM regulator are determined by the voltage reference circuit, dynamic voltage characteristics are dictated by the power stages. The power stage is therefore a critical element in overall DRAM performance. To date, the most prevalent type of power stage among DRAM designers is a simple, unbuffered op-amp. Unbuffered op-amps, while providing high open loop gain, fast response, and low offset, allow design engineers to use feedback in the overall regulator design. Feedback reduces temperature and process sensitivity and ensures better load regulation than any type of open loop system. Design of the op-amps, however, is anything but simple.

The ideal power stage would have high bandwidth, high open-loop gain, high slew rate, low systematic offset, low operating current, high drive, and inherent stability. Unfortunately, several of these parameters are contradictory, which compromises certain aspects of the design and necessitates

Sec. 6.1 Internal Voltage Regulators

trade-offs. While it seems that many DRAM manufacturers use a single op-amp for the regulator's power stage, we have found that it is better to use a multitude of smaller op-amps. These smaller op-amps have wider bandwidth, greater design flexibility, and an easier layout than a single, large op-amp.

The power op-amp is shown in Figure 6.7. The schematic diagram for a voltage regulator power stage is shown in Figure 6.8. This design is used on a 256Mb DRAM and consists of 18 power op-amps, one boost amp, and one small standby op-amp. The V_{CC} power buses for the array and peripheral circuits are isolated except for the 20-ohm resistor that bridges the two together. Isolating the buses is important to prevent high-current spikes that occur in the array circuits from affecting the peripheral circuits. Failure to isolate these buses can result in speed degradation for the DRAM because high-current spikes in the array cause voltage cratering and a corresponding slowdown in logic transitions.

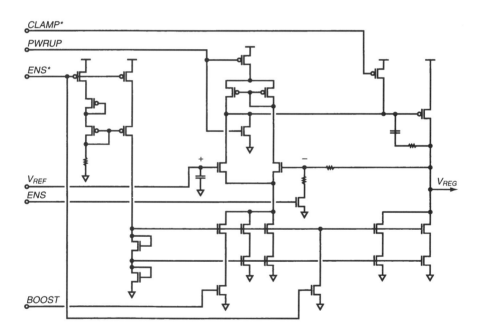

Figure 6.7 Power op-amp.

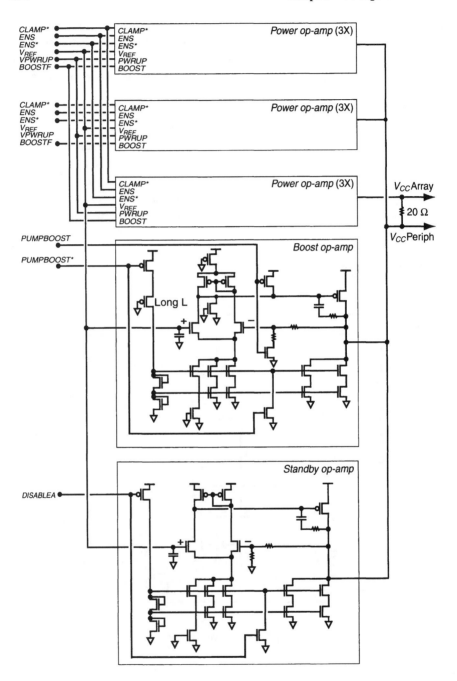

Figure 6.8 Power stage.

With isolation, the peripheral V_{CC} is almost immune to array noise. To improve slew rate, each of the power op-amp stages shown in Figure 6.8 features a boost circuit that raises the differential pair bias current and slew rate during expected periods of large current spikes. Large spikes are normally associated with Psense-amp activation. To reduce active current consumption, the boost current is disabled a short time after Psense-amp activation by the signal labeled *BOOST*. The power stages themselves are enabled by the signals labeled *ENS* and *ENS** only when $\overline{RAS}$ is LOW and the part is active. When $\overline{RAS}$ is HIGH, all of the power stages are disabled.

The signal labeled *CLAMP** ensures that the PMOS output transistor is OFF whenever the amplifier is disabled to prevent unwanted charging of the V_{CC} bus. When forced to ground, however, the signal labeled *PWRUP* shorts the V_{CCX} and V_{CC} buses together through the PMOS output transistor. (The need for this function was described earlier in this section.) Basically, these two buses are shorted together whenever the DRAM operates in Region 1, as depicted in Figure 6.1. Obviously, to prevent a short circuit between V_{CCX} and ground, *CLAMP** and *PWRUP* are mutually exclusive.

The smaller standby amplifier is included in this design to sustain the V_{CC} supply whenever the part is inactive, as determined by $\overline{RAS}$. This amplifier has a very low operating current and a correspondingly low slew rate. Accordingly, the standby amplifier cannot sustain any type of active load. For this reason, the third and final type of amplifier is included. The boost amplifier shown in Figure 6.8 is identical to the power stage amplifiers except that the bias current is greatly reduced. This amplifier provides the necessary supply current to operate the V_{CCP} and V_{BB} voltage pumps.

The final element in the voltage regulator is the control logic. An example of this logic is shown in Figure 6.9. It consists primarily of static CMOS logic gates and level translators. The logic gates are referenced to V_{CC}. Level translators are necessary to drive the power stages, which are referenced to V_{CCX} levels. A series of delay elements tune the control circuit relative to *Psense-amp activation (ACT)* and $\overline{RAS}$ *(RL*)* timing. Included in the control circuit is the block labeled V_{CCX} *level detector* [1]. The *reference generator* generates two reference signals, which are fed into the comparator, to determine the transition point V1 between Region 1 and Region 2 operation for the regulator. In addition, the *boost amp control* logic block is shown in Figure 6.9. This circuit examines the V_{BB} and V_{CCP} control signals to enable the boost amplifier whenever either voltage pump is active.

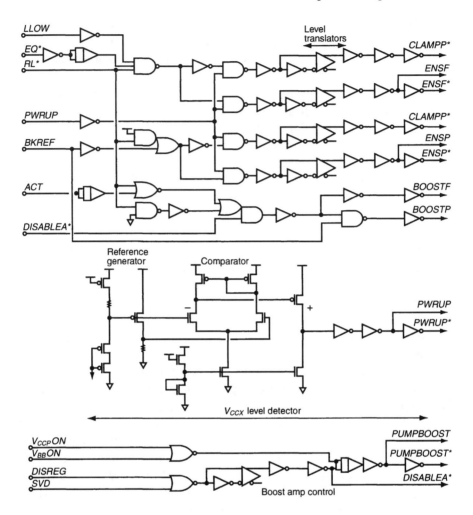

Figure 6.9 Regulator control logic.

6.2 PUMPS AND GENERATORS

Generation of the boosted wordline voltage and negative substrate bias voltage requires voltage pump (or charge pump) circuits. Pump circuits are commonly used to create voltages that are more positive or negative than available supply voltages. Two voltage pumps are commonly used in a DRAM today. The first is a V_{CCP} pump, which generates the boosted wordline voltage and is built primarily from NMOS transistors. The second is a V_{BB} pump, which generates the negative substrate bias voltage and is built from PMOS transistors. The exclusive use of NMOS or PMOS transistors in each pump is required to prevent latch-up and current injection into the mbit arrays. NMOS transistors are required in the V_{CCP} pump because various active nodes would swing negative with respect to the substrate voltage V_{CCP}. Any n-diffusion regions connected to these active nodes would forward bias and cause latch-up and injection. Similar conditions mandate the use of PMOS transistors in the V_{BB} pump.

6.2.1 Pumps

Voltage pump operation can be understood with the assistance of the simple voltage pump circuit depicted in Figure 6.10. For this positive pump circuit, imagine, for one phase of a pump cycle, that the clock *CLK* is HIGH. During this phase, node A is at ground and node B is clamped to $V_{CC}-V_{TH}$ by transistor M1. The charge stored in capacitor C1 is then

$$Q1 = C1 \cdot (V_{CC} - V_{TH}) \text{ coulombs.}$$

During the second phase, the clock *CLK* will transition LOW, which brings node A HIGH. As node A rises to V_{CC}, node B begins to rise to $V_{CC} + (V_{CC}-V_{TH})$, shutting OFF transistor M1. At the same time, as node B rises one V_{TH} above V_{LOAD}, transistor M2 begins to conduct. The charge from capacitor C1 is transferred through M2 and shared with the capacitor C_{LOAD}. This action effectively pumps charge into C_{LOAD} and ultimately raises the voltage V_{OUT}. During subsequent clock cycles, the voltage pump continues to deliver charge to C_{LOAD} until the voltage V_{OUT} equals $2V_{CC}-V_{TH1}-V_{TH2}$, one V_{TH} below the peak voltage occurring at node B. A simple, negative voltage pump could be built from the circuit of Figure 6.10 by substituting PMOS transistors for the two NMOS transistors shown and moving their respective gate connections.

Schematics for actual V_{CCP} and V_{BB} pumps are shown in Figures 6.11 and 6.12, respectively. Both of these circuits are identical except for the changes associated with the NMOS and PMOS transistors. These pump circuits operate as two phase pumps because two identical pumps are working in tandem. As discussed in the previous paragraph, note that transistors M1 and M2 are configured as switches rather than as diodes. The drive signals for these gates are derived from secondary pump stages and the tandem pump circuit. Using switches rather than diodes improves pumping efficiency and operating range by eliminating the V_{TH} drops associated with diodes.

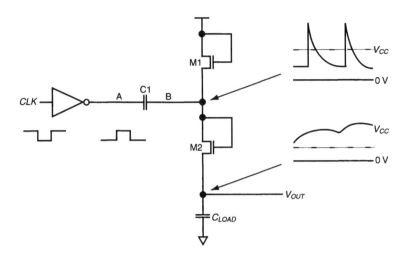

Figure 6.10 Simple voltage pump circuit.

Two important characteristics of a voltage pump are capacity and efficiency. Capacity is a measure of how much current a pump can continue to supply, and it is determined primarily by the capacitor's size and its operating frequency. The operating frequency is limited by the rate at which the pump capacitor C1 can be charged and discharged. Efficiency, on the other hand, is a measure of how much charge or current is wasted during each pump cycle. A typical DRAM voltage pump might be 30–50% efficient. This translates into 2–3 milliamps of supply current for every milliamp of pump output current.

In addition to the pump circuits just described, regulator and oscillator circuits are needed to complete a voltage pump design. The most common oscillator used in voltage pumps is the standard CMOS ring oscillator. A typical voltage pump ring oscillator is shown in Figure 6.13. A unique feature of this oscillator is the multifrequency operation permitted by including mux circuits connected to various oscillator tap points. These muxes, controlled by signals such as *PWRUP*, enable higher frequency operation by reducing the number of inverter stages in the ring oscillator.

Typically, the oscillator is operated at a higher frequency when the DRAM is in a *PWRUP* state, since this will assist the pump in initially charging the load capacitors. The oscillator is enabled and disabled through the signal labeled *REGDIS**. This signal is controlled by the voltage regulator circuit shown in Figure 6.14. Whenever *REGDIS** is HIGH, the oscillator is functional, and the pump is operative. Examples of V_{CCP} and V_{BB} pump regulators are shown in Figure 6.14 and 6.15, respectively.

Sec. 6.2 Pumps and Generators

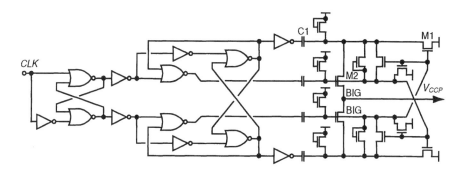

Figure 6.11 V_{CCP} pump.

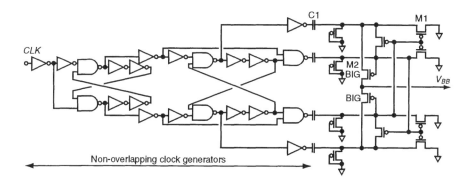

Figure 6.12 V_{BB} pump.

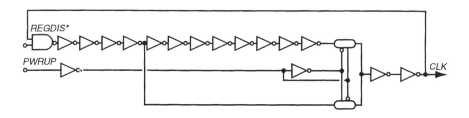

Figure 6.13 Ring oscillator.

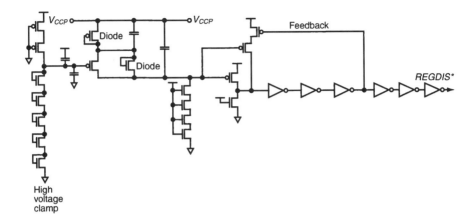

Figure 6.14 V_{CCP} regulator.

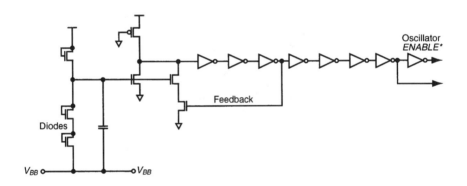

Figure 6.15 V_{BB} regulator.

These circuits use low-bias current sources and NMOS/PMOS diodes to translate the pumped voltage levels, V_{CCP} and V_{BB}, to normal voltage levels. This shifted voltage level is fed into a modified inverter stage that has an output-dependent trip point (the inverter has hysteresis). The trip point is modified with feedback to provide hysteresis for the circuit. Subsequent inverter stages provide additional gain for the regulator and boost the signal to the full CMOS level necessary to drive the oscillator. Minimum and maximum operating voltages for the pump are controlled by the first inverter stage trip point, hysteresis, and the NMOS/PMOS diode voltages. The clamp circuit shown in Figure 6.14 is included in the regulator design to limit the pump voltage when V_{CC} is elevated, such as during burn-in.

A second style of voltage pump regulator for controlling V_{CCP} and V_{BB} voltages is shown in Figure 6.16 and 6.17, respectively. This type of regulator uses a high-gain comparator coupled to a voltage translator stage. The translator stage of Figure 6.16 translates the pumped voltage V_{CCP} and the reference voltage V_{DD} down within the input common-mode range of the comparator circuit. The translator accomplishes this with a reference current source and MOS diodes. The reference voltage supply V_{DD} is translated down by one threshold voltage (V_{TH}) by sinking the reference current with a current mirror stage through a PMOS diode connected to V_{DD}. The pumped voltage supply V_{CCP} is similarly translated down by sinking the same reference current with a matching current mirror stage through a diode stack. The diode stack consists of a PMOS diode, matching that in the V_{DD} reference translator, and a pseudo-NMOS diode. The pseudo-NMOS diode is actually a series of NMOS transistors with a common gate connection.

The quantity and sizes of the transistors included in this pseudo-NMOS diode are mask-programmable. The voltage drop across this pseudo-NMOS diode determines, in essence, the regulated voltage for V_{CCP}. Accordingly,

$$V_{CCP} = V_{DD} + Vndiode.$$

The voltage dropped across the PMOS diode does not affect the regulated voltage because the reference voltage supply V_{DD} is translated through a matching PMOS diode. Both of the translated voltages are fed into the comparator stage, which enables the pump oscillator whenever the translated V_{CCP} voltage falls below the translated V_{DD} reference voltage. The comparator has built-in hysteresis, via the middle stage: this dictates the amount of ripple present on the regulated V_{CCP} supply.

The V_{BB} regulator in Figure 6.17 operates in a similar fashion to the V_{CCP} regulator of Figure 6.16. The primary difference lies in the voltage translator stage. For the V_{BB} regulator, this stage translates the pumped voltage V_{BB} and the reference voltage V_{SS} up within the input common mode range of the comparator circuit. The reference voltage V_{SS} is translated up by one threshold voltage (V_{TH}) by sourcing a reference current with a current mirror stage through an NMOS diode. The regulated voltage V_{BB} is similarly translated up by sourcing the same reference current with a matching current mirror stage through a diode stack. This diode stack, similar to the V_{CCP} case, contains an NMOS diode that matches that used in translating the reference voltage V_{SS}. The stack also contains a mask-adjustable, pseudo-NMOS diode. The voltage across the pseudo-NMOS diode determines the regulated voltage for V_{BB} such that

$$V_{BB} = -Vndiode.$$

The comparator includes a hysteresis stage, which dictates the amount of ripple present on the regulated V_{BB} supply.

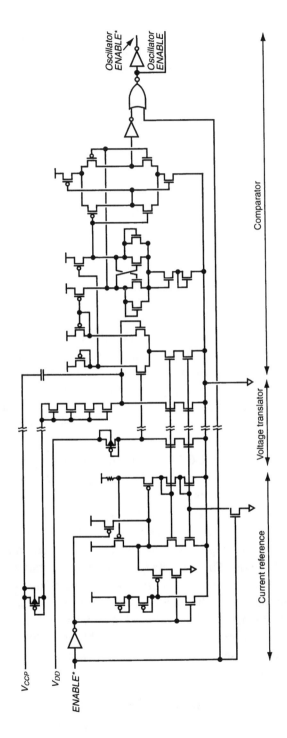

Figure 6.16 V_{CCP} differential regulator.

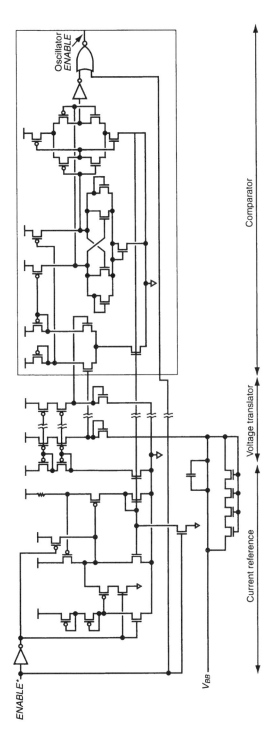

Figure 6.17 V_{BB} differential regulator.

6.2.2 *DVC*2 Generator

With our discussion of voltage regulator circuits concluded, we can briefly turn our attention to the *DVC*2 generation. As discussed in Section 1.2, the memory capacitor cellplate is biased to $V_{CC}/2$. Furthermore, the digitlines are always equilibrated and biased to $V_{CC}/2$ between array accesses. In most DRAM designs, the cellplate and digitline bias voltages are derived from the same generator circuit. A simple circuit for generating $V_{CC}/2$ (*DVC*2) voltage is shown in Figure 6.18. It consists of a standard CMOS inverter with the input and output terminals shorted together. With correct transistor sizing, the output voltage of this circuit can be set precisely to $V_{CC}/2$ V. Taking this simple design one step further results in the actual DRAM *DVC*2 generator found in Figure 6.19. This generator contains additional transistors to improve both its stability and drive capability.

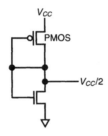

Figure 6.18 Simple *DVC*2 generator.

6.3 DISCUSSION

In this chapter, we introduced the popular circuits used on a DRAM for voltage generation and regulation. Because this introduction is far from exhaustive, we include a list of relevant readings and references in the Appendix for those readers interested in greater detail.

Sec. 6.3 Discussion   175

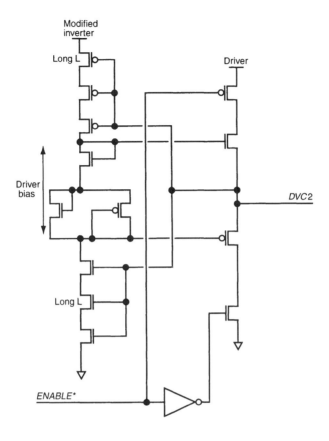

Figure 6.19 *DVC*2 generator.

REFERENCES

[1] R. J. Baker, H. W. Li, and D. E. Boyce, *CMOS: Circuit Design, Layout, and Simulation.* Piscataway, NJ: IEEE Press, 1998.

[2] B. Keeth, *Control Circuit Responsive to Its Supply Voltage Level,* United States Patent #5,373,227, December 13, 1994.

Appendix

Supplemental Reading

In this tutorial overview of DRAM circuit design, we may not have covered specific topics to the reader's satisfaction. For this reason, we have compiled a list of supplemental readings from major conferences and journals, categorized by subject. It is our hope that unanswered questions will be addressed by the authors of these readings, who are experts in the field of DRAM circuit design.

General DRAM Design and Operation

[1] S. Fuji, K. Natori, T. Furuyama, S. Saito, H. Toda, T. Tanaka, and O. Ozawa, "A Low-Power Sub 100 ns 256K Bit Dynamic RAM," *IEEE Journal of Solid-State Circuits*, vol. 18, pp. 441–446, October 1983.

[2] A. Mohsen, R. I. Kung, C. J. Simonsen, J. Schutz, P. D. Madland, E. Z. Hamdy, and M. T. Bohr, "The Design and Performance of CMOS 256K Bit DRAM Devices," *IEEE Journal of Solid-State Circuits*, vol. 19, pp. 610–618, October 1984.

[3] M. Aoki, Y. Nakagome, M. Horiguchi, H. Tanaka, S. Ikenaga, J. Etoh, Y. Kawamoto, S. Kimura, E. Takeda, H. Sunami, and K. Itoh, "A 60-ns 16-Mbit CMOS DRAM with a Transposed Data-Line Structure," *IEEE Journal of Solid-State Circuits*, vol. 23, pp. 1113–1119, October 1988.

[4] M. Inoue, T. Yamada, H. Kotani, H. Yamauchi, A. Fujiwara, J. Matsushima, H. Akamatsu, M. Fukumoto, M. Kubota, I. Nakao, N. Aoi, G. Fuse, S. Ogawa, S. Odanaka, A. Ueno, and H. Yamamoto, "A 16-Mbit DRAM with a Relaxed Sense-Amplifier-Pitch Open-Bit-Line Architecture," *IEEE Journal of Solid-State Circuits*, vol. 23, pp. 1104–1112, October 1988.

[5] T. Watanabe, G. Kitsukawa, Y. Kawajiri, K. Itoh, R. Hori, Y. Ouchi, T. Kawahara, and R. Matsumoto, "Comparison of CMOS and BiCMOS 1-Mbit DRAM Performance," *IEEE Journal of Solid-State Circuits*, vol. 24, pp. 771–778, June 1989.

[6] K. Itoh, "Trends in Megabit DRAM Circuit Design," *IEEE Journal of Solid-State Circuits*, vol. 25, pp. 778–789, June 1990.

[7] Y. Nakagome, H. Tanaka, K. Takeuchi, E. Kume, Y. Watanabe, T. Kaga, Y. Kawamoto, F. Murai, R. Izawa, D. Hisamoto, T. Kisu, T. Nishida, E. Takeda, and K. Itoh, "An Experimental 1.5-V 64-Mb DRAM," *IEEE Journal of Solid-State Circuits*, vol. 26, pp. 465–472, April 1991.

[8] P. Gillingham, R. C. Foss, V. Lines, G. Shimokura, and T. Wojcicki, "High-Speed, High-Reliability Circuit Design for Megabit DRAM," *IEEE Journal of Solid-State Circuits*, vol. 26, pp. 1171–1175, August 1991.

[9] K. Kimura, T. Sakata, K. Itoh, T. Kaga, T. Nishida, and Y. Kawamoto, "A Block-Oriented RAM with Half-Sized DRAM Cell and Quasi-Folded Data-Line Architecture," *IEEE Journal of Solid-State Circuits*, vol. 26, pp. 1511–1518, November 1991.

[10] Y. Oowaki, K. Tsuchida, Y. Watanabe, D. Takashima, M. Ohta, H. Nakano, S. Watanabe, A. Nitayama, F. Horiguchi, K. Ohuchi, and F. Masuoka, "A 33-ns 64-Mb DRAM," *IEEE Journal of Solid-State Circuits*, vol. 26, pp. 1498–1505, November 1991.

[11] T. Kirihata, S. H. Dhong, K. Kitamura, T. Sunaga, Y. Katayama, R. E. Scheuerlein, A. Satoh, Y. Sakaue, K. Tobimatus, K. Hosokawa, T. Saitoh, T. Yoshikawa, H. Hashimoto, and M. Kazusawa, "A 14-ns 4-Mb CMOS DRAM with 300-mW Active Power," *IEEE Journal of Solid-State Circuits*, vol. 27, pp. 1222–1228, September 1992.

[12] K. Shimohigashi and K. Seki, "Low-Voltage ULSI Design," *IEEE Journal of Solid-State Circuits*, vol. 28, pp. 408–413, April 1993.

[13] G. Kitsukawa, M. Horiguchi, Y. Kawajiri, T. Kawahara, T. Akiba, Y. Kawase, T. Tachibana, T. Sakai, M. Aoki, S. Shukuri, K. Sagara, R. Nagai, Y. Ohji, N. Hasegawa, N. Yokoyama, T. Kisu, H. Yamashita, T. Kure, and T. Nishida, "256-Mb DRAM Circuit Technologies for File Applications," *IEEE Journal of Solid-State Circuits*, vol. 28, pp. 1105–1113, November 1993.

[14] T. Kawahara, Y. Kawajiri, M. Horiguchi, T. Akiba, G. Kitsukawa, T. Kure, and M. Aoki, "A Charge Recycle Refresh for Gb-Scale DRAM's in File Applications," *IEEE Journal of Solid-State Circuits*, vol. 29, pp. 715–722, June 1994.

[15] S. Shiratake, D. Takashima, T. Hasegawa, H. Nakano, Y. Oowaki, S. Watanabe, K. Ohuchi, and F. Masuoka, "A Staggered NAND DRAM Array Architecture for a Gbit Scale Integration," 1994 Symposium on VLSI Circuits, p. 75, June 1994.

[16] T. Ooishi, K. Hamade, M. Asakura, K. Yasuda, H. Hidaka, H. Miyamoto, and H. Ozaki, "An Automatic Temperature Compensation of Internal Sense Ground for Sub-Quarter Micron DRAMs," 1994 Symposium on VLSI Circuits, p. 77, June 1994.

[17] A. Fujiwara, H. Kikukawa, K. Matsuyama, M. Agata, S. Iwanari, M. Fukumoto, T. Yamada, S. Okada, and T. Fujita, "A 200MHz 16Mbit

Synchronous DRAM with Block Access Mode," 1994 Symposium on VLSI Circuits, p. 79, June 1994.

[18] Y. Kodama, M. Yanagisawa, K. Shigenobu, T. Suzuki, H. Mochizuki, and T. Ema, "A 150-MHz 4-Bank 64M-bit SDRAM with Address Incrementing Pipeline Scheme," 1994 Symposium on VLSI Circuits, p. 81, June 1994.

[19] D. Choi, Y. Kim, G. Cha, J. Lee, S. Lee, K. Kim, E. Haq, D. Jun, K. Lee, S. Cho, J. Park, and H. Lim, "Battery Operated 16M DRAM with Post Package Programmable and Variable Self Refresh," 1994 Symposium on VLSI Circuits, p. 83, June 1994.

[20] S. Yoo, J. Han, E. Haq, S. Yoon, S. Jeong, B. Kim, J. Lee, T. Jang, H. Kim, C. Park, D. Seo, C. Choi, S. Cho, and C. Hwang, "A 256M DRAM with Simplified Register Control for Low Power Self Refresh and Rapid Burn-In," 1994 Symposium on VLSI Circuits, p. 85, June 1994.

[21] M. Tsukude, M. Hirose, S. Tomishima, T. Tsuruda, T. Yamagata, K. Arimoto, and K. Fujishima, "Automatic Voltage-Swing Reduction (AVR) Scheme for Ultra Low Power DRAMs," 1994 Symposium on VLSI Circuits, p. 87, June 1994.

[22] D. Stark, H. Watanabe, and T. Furuyama, "An Experimental Cascade Cell Dynamic Memory," 1994 Symposium on VLSI Circuits, p. 89, June 1994.

[23] T. Inaba, D. Takashima, Y. Oowaki, T. Ozaki, S. Watanabe, and K. Ohuchi, "A 250mV Bit-Line Swing Scheme for a 1V 4Gb DRAM," 1995 Symposium on VLSI Circuits, p. 99, June 1995.

[24] I. Naritake, T. Sugibayashi, S. Utsugi, and T. Murotani, "A Crossing Charge Recycle Refresh Scheme with a Separated Driver Sense-Amplifier for Gb DRAMs," 1995 Symposium on VLSI Circuits, p. 101, June 1995.

[25] S. Kuge, T. Tsuruda, S. Tomishima, M. Tsukude, T. Yamagata, and K. Arimoto, "SOI-DRAM Circuit Technologies for Low Power High Speed Multi-Giga Scale Memories," 1995 Symposium on VLSI Circuits, p. 103, June 1995.

[26] Y. Watanabe, H. Wong, T. Kirihata, D. Kato, J. DeBrosse, T. Hara, M. Yoshida, H. Mukai, K. Quader, T. Nagai, P. Poechmueller, K. Pfefferl, M. Wordeman, and S. Fujii, "A 286mm^2 256Mb DRAM with X32 Both-Ends DQ," 1995 Symposium on VLSI Circuits, p. 105, June 1995.

[27] T. Kirihata, Y. Watanabe, H. Wong, J. DeBrosse, M. Yoshida, D. Katoh, S. Fujii, M. Wordeman, P. Poechmueller, S. Parke, and Y. Asao, "Fault-Tolerant Designs for 256 Mb DRAM," 1995 Symposium on VLSI Circuits, p. 107, June 1995.

[28] D. Takashima, Y. Oowaki, S. Watanabe, and K. Ohuchi, "A Novel Power-Off Mode for a Battery-Backup DRAM," 1995 Symposium on VLSI Circuits, p. 109, June 1995.

[29] T. Ooishi, Y. Komiya, K. Hamada, M. Asakura, K. Yasuda, K. Furutani, T. Kato, H. Hidaka, and H. Ozaki, "A Mixed-Mode Voltage-Down Converter

with Impedance Adjustment Circuitry for Low-Voltage Wide-Frequency DRAMs," 1995 Symposium on VLSI Circuits, p. 111, June 1995.

[30] S.-J. Lee, K.-W. Park, C.-H. Chung, J.-S. Son, K.-H. Park, S.-H. Shin, S. T. Kim, J.-D. Han, H.-J. Yoo, W.-S. Min, and K.-H. Oh, "A Low Noise 32Bit-Wide 256M Synchronous DRAM with Column-Decoded I/O Line," 1995 Symposium on VLSI Circuits, p. 113, June 1995.

[31] T. Sugibayashi, I. Naritake, S. Utsugi, K. Shibahara, R. Oikawa, H. Mori, S. Iwao, T. Murotani, K. Koyama, S. Fukuzawa, T. Itani, K. Kasama, T. Okuda, S. Ohya, and M. Ogawa, "A 1-Gb DRAM for File Applications," *IEEE Journal of Solid-State Circuits*, vol. 30, pp. 1277–1280, November 1995.

[32] T. Yamagata, S. Tomishima, M. Tsukude, T. Tsuruda, Y. Hashizume, and K. Arimoto, "Low Voltage Circuit Design Techniques for Battery-Operated and/or Giga-Scale DRAMs," *IEEE Journal of Solid-State Circuits*, vol. 30, pp. 1183–1188, November 1995.

[33] H. Nakano, D. Takashima, K. Tsuchida, S. Shiratake, T. Inaba, M. Ohta, Y. Oowaki, S. Watanabe, K. Ohuchi, and J. Matsunaga, "A Dual Layer Bitline DRAM Array with V_{CC}/V_{SS} Hybrid Precharge for Multi-Gigabit DRAMs," 1996 Symposium on VLSI Circuits, p. 190, June 1996.

[34] J. Han, J. Lee, S. Yoon, S. Jeong, C. Park, I. Cho, S. Lee, and D. Seo, "Skew Minimization Techniques for 256M-bit Synchronous DRAM and Beyond," 1996 Symposium on VLSI Circuits, p. 192, June 1996.

[35] H. Wong, T. Kirihata, J. DeBrosse, Y. Watanabe, T. Hara, M. Yoshida, M. Wordeman, S. Fujii, Y. Asao, and B. Krsnik, "Flexible Test Mode Design for DRAM Characterization," 1996 Symposium on VLSI Circuits, p. 194, June 1996.

[36] D. Takashima, Y. Oowaki, S. Watanabe, K. Ohuchi, and J. Matsunaga, "Noise Suppression Scheme for Giga-Scale DRAM with Hundreds of I/Os," 1996 Symposium on VLSI Circuits, p. 196, June 1996.

[37] S. Tomishima, F. Morishita, M. Tsukude, T. Yamagata, and K. Arimoto, "A Long Data Retention SOI-DRAM with the Body Refresh Function," 1996 Symposium on VLSI Circuits, p. 198, June 1996.

[38] M. Nakamura, T. Takahashi, T. Akiba, G. Kitsukawa, M. Morino, T. Sekiguchi, I. Asano, K. Komatsuzaki, Y. Tadaki, Songsu Cho, K. Kajigaya, T. Tachibana, and K. Sato, "A 29-ns 64-Mb DRAM with Hierarchical Array Architecture," *IEEE Journal of Solid-State Circuits*, vol. 31, pp. 1302–1307, September 1996.

[39] K. Itoh, Y. Nakagome, S. Kimura, and T. Watanabe, "Limitations and Challenges of Multigigabit DRAM Chip Design," *IEEE Journal of Solid-State Circuits*, vol. 32, pp. 624–634, May 1997.

[40] Y. Idei, K. Shimohigashi, M. Aoki, H. Noda, H. Iwai, K. Sato, and T. Tachibana, "Dual-Period Self-Refresh Scheme for Low-Power DRAM's with On-Chip PROM Mode Register," *IEEE Journal of Solid-State Circuits*, vol. 33, pp. 253–259, February 1998.

[41] K. Kim, C.-G. Hwang, and J. G. Lee, "DRAM Technology Perspective for Gigabit Era," *IEEE Transactions Electron Devices*, vol. 45, pp. 598–608, March 1998.

[42] H. Tanaka, M. Aoki, T. Sakata, S. Kimura, N. Sakashita, H. Hidaka, T. Tachibana, and K. Kimura, "A Precise On-Chip Voltage Generator for a Giga-Scale DRAM with a Negative Word-Line Scheme," 1998 Symposium on VLSI Circuits, p. 94, June 1998.

[43] T. Fujino and K. Arimoto, "Multi-Gbit-Scale Partially Frozen (PF) NAND DRAM with SDRAM Compatible Interface," 1998 Symposium on VLSI Circuits, p. 96, June 1998.

[44] A. Yamazaki, T. Yamagata, M. Hatakenaka, A. Miyanishi, I. Hayashi, S. Tomishima, A. Mangyo, Y. Yukinari, T. Tatsumi, M. Matsumura, K. Arimoto, and M. Yamada, "A 5.3Gb/s 32Mb Embedded SDRAM Core with Slightly Boosting Scheme," 1998 Symposium on VLSI Circuits, p. 100, June 1998.

[45] C. Kim, K. H. Kyung, W. P. Jeong, J. S. Kim, B. S. Moon, S. M. Yim, J. W. Chai, J. H. Choi, C. K. Lee, K. H. Han, C. J. Park, H. Choi, and S. I. Cho, "A 2.5V, 2.0GByte/s Packet-Based SDRAM with a 1.0Gbps/pin Interface," 1998 Symposium on VLSI Circuits, p. 104, June 1993.

[46] M. Saito, J. Ogawa, H. Tamura, S. Wakayama, H. Araki, Tsz-Shing Cheung, K. Gotoh, T. Aikawa, T. Suzuki, M. Taguchi, and T. Imamura, "500-Mb/s Nonprecharged Data Bus for High-Speed Dram's," *IEEE Journal of Solid-State Circuits*, vol. 33, pp. 1720–1730, November 1998.

DRAM Cells

[47] C. G. Sodini and T. I. Kamins, "Enhanced Capacitor for One-Transistor Memory Cell," *IEEE Transactions Electron Devices*, vol. ED-23, pp. 1185–1187, October 1976.

[48] J. E. Leiss, P. K. Chatterjee, and T. C. Holloway, "DRAM Design Using the Taper-Isolated Dynamic RAM Cell," *IEEE Journal of Solid-State Circuits*, vol. 17, pp. 337–344, April 1982.

[49] K. Yamaguchi, R. Nishimura, T. Hagiwara, and H. Sunami, "Two-Dimensional Numerical Model of Memory Devices with a Corrugated Capacitor Cell Structure," *IEEE Journal of Solid-State Circuits*, vol. 20, pp. 202–209, February 1985.

[50] N. C. Lu, P. E. Cottrell, W. J. Craig, S. Dash, D. L. Critchlow, R. L. Mohler, B. J. Machesney, T. H. Ning, W. P. Noble, R. M. Parent, R. E. Scheuerlein, E. J. Sprogis, and L. M. Terman, "A Substrate-Plate Trench-Capacitor (SPT) Memory Cell for Dynamic RAM's," *IEEE Journal of Solid-State Circuits*, vol. 21, pp. 627–634, October 1986.

[51] Y. Nakagome, M. Aoki, S. Ikenaga, M. Horiguchi, S. Kimura, Y. Kawamoto, and K. Itoh, "The Impact of Data-Line Interference Noise on DRAM

Scaling," *IEEE Journal of Solid-State Circuits*, vol. 23, pp. 1120–1127, October 1988.

[52] K. W. Kwon. I. S. Park, D. H. Han, E. S. Kim, S. T. Ahn, and M. Y. Lee, "Ta_2O_5 Capacitors for 1 Gbit DRAM and Beyond," *1994 IEDM Technical Digest*, pp. 835–838.

[53] D. Takashima, S. Watanabe, H. Nakano, Y. Oowaki, and K. Ohuchi, "Open/Folded Bit-Line Arrangement for Ultra-High-Density DRAM's," *IEEE Journal of Solid-State Circuits*, vol. 29, pp. 539–542, April 1994.

[54] Wonchan Kim, Joongsik Kih, Gyudong Kim, Sanghun Jung, and Gijung Ahn, "An Experimental High-Density DRAM Cell with a Built-in Gain Stage," *IEEE Journal of Solid-State Circuits*, vol. 29, pp. 978–981, August 1994.

[55] Y. Kohyama, T. Ozaki, S. Yoshida, Y. Ishibashi, H. Nitta, S. Inoue, K. Nakamura, T. Aoyama, K. Imai, and N. Hayasaka, "A Fully Printable, Self-Aligned and Planarized Stacked Capacitor DRAM Access Cell Technology for 1 Gbit DRAM and Beyond," *IEEE Symposium on VLSI Technology Digest of Technical Papers*, pp. 17–18, June 1997.

[56] B. El-Kareh, G. B. Bronner, and S. E. Schuster, "The Evolution of DRAM Cell Technology, *Solid State Technology*, vol. 40, pp. 89–101, May 1997.

[57] S. Takehiro, S. Yamauchi, M. Yoshimaru, and H. Onoda, "The Simplest Stacked BST Capacitor for Future DRAM's Using a Novel Low Temperature Growth Enhanced Crystallization," *IEEE Symposium on VLSI Technology Digest of Technical Papers*, pp. 153–154, June 1997.

[58] T. Okuda and T. Murotani, "A Four-Level Storage 4-Gb DRAM," *IEEE Journal of Solid State Circuits*, vol. 32, pp. 1743–1747, November 1997.

[59] A. Nitayama, Y. Kohyama, and K. Hieda, "Future Directions for DRAM Memory Cell Technology," *1998 IEDM Technical Digest*, pp. 355–358.

[60] K. Ono, T. Horikawa, T. Shibano, N. Mikami, T. Kuroiwa, T. Kawahara, S. Matsuno, F. Uchikawa, S. Satoh, and H. Abe, "(Ba, Sr)TiO_3 Capacitor Technology for Gbit-Scale DRAMs," *1998 IEDM Technical Digest*, pp. 803–806.

[61] H. J. Levy, E. S. Daniel, and T. C. McGill, "A Transistorless-Current-Mode Static RAM Architecture," *IEEE Journal of Solid-State Circuits*, vol. 33, pp. 669–672, April 1998.

DRAM Sensing

[62] N. C.-C. Lu and H. H. Chao, "Half-V_{DD}/Bit-Line Sensing Scheme in CMOS DRAMs," *IEEE Journal of Solid-State Circuits*, vol. 19, pp. 451–454, August 1984.

[63] P. A. Layman and S. G. Chamberlain, "A Compact Thermal Noise Model for the Investigation of Soft Error Rates in MOS VLSI Digital Circuits," *IEEE Journal of Solid-State Circuits*, vol. 24, pp. 79–89, February 1989.

[64] R. Kraus, "Analysis and Reduction of Sense-Amplifier Offset," *IEEE Journal of Solid-State Circuits*, vol. 24, pp. 1028–1033, August 1989.

[65] R. Kraus and K. Hoffmann, "Optimized Sensing Scheme of DRAMs," *IEEE Journal of Solid-State Circuits*, vol. 24, pp. 895–899, August 1989.

[66] H. Hidaka, Y. Matsuda, and K. Fujishima, "A Divided/Shared Bit-Line Sensing Scheme for ULSI DRAM Cores," *IEEE Journal of Solid-State Circuits*, vol. 26, pp. 473–478, April 1991.

[67] T. Nagai, K. Numata, M. Ogihara, M. Shimizu, K. Imai, T. Hara, M. Yoshida, Y. Saito, Y. Asao, S. Sawada, and S. Fuji, "A 17-ns 4-Mb CMOS DRAM," *IEEE Journal of Solid-State Circuits*, vol 26, pp. 1538–1543, November 1991.

[68] T. N. Blalock and R. C. Jaeger, "A High-Speed Sensing Scheme for 1T Dynamic RAMs Utilizing the Clamped Bit-Line Sense Amplifier," *IEEE Journal of Solid-State Circuits*, vol. 27, pp. 618–625, April 1992.

[69] M. Asakura, T. Ooishi, M. Tsukude, S. Tomishima, T. Eimori, H. Hidaka, Y. Ohno, K. Arimoto, K. Fujishima, T. Nishimura, and T. Yoshihara, "An Experimental 256-Mb DRAM with Boosted Sense-Ground Scheme," *IEEE Journal of Solid-State Circuits*, vol. 29, pp. 1303–1309, November 1994.

[70] T. Eirihata, S. H. Dhong, L. M. Terman, T. Sunaga, and Y. Taira, "A Variable Precharge Voltage Sensing," *IEEE Journal of Solid-State Circuits*, vol. 30, pp. 25–28, January 1995.

[71] T. Hamamoto, Y. Morooka, M. Asakura, and H. Ozaki, "Cell-Plate-Line/Bit-Line Complementary Sensing (CBCS) Architecture for Ultra Low-Power DRAMs," *IEEE Journal of Solid-State Circuits*, vol. 31, pp. 592–601, April 1996.

[72] T. Sunaga, "A Full Bit Prefetch DRAM Sensing Circuit," *IEEE Journal of Solid-State Circuits*, vol. 31, pp. 767–772, June 1996.

DRAM On-Chip Voltage Generation

[73] M. Horiguchi, M. Aoki, J. Etoh, H. Tanaka, S. Ikenaga, K. Itoh, K. Kajigaya, H. Kotani, K. Ohshima, and T. Matsumoto, "A Tunable CMOS-DRAM Voltage Limiter with Stabilized Feedback Amplifier," *IEEE Journal of Solid-State Circuits*, vol. 25, pp. 1129–1135, October 1990.

[74] D. Takashima, S. Watanabe, T. Fuse, K. Sunouchi, and T. Hara, "Low-Power On-Chip Supply Voltage Conversion Scheme for Ultrahigh-Density

DRAMs," *IEEE Journal of Solid-State Circuits*, vol. 28, pp. 504–509, April 1993.

[75] T. Kuroda, K. Suzuki, S. Mita, T. Fujita, F. Yamane, F. Sano, A. Chiba, Y. Watanabe, K. Matsuda, T. Maeda, T. Sakurai, and T. Furuyama, "Variable Supply-Voltage Scheme for Low-Power High-Speed CMOS Digital Design," *IEEE Journal of Solid-State Circuits*, vol. 33, pp. 454–462, March 1998.

DRAM SOI

[76] S. Kuge, F. Morishita, T. Tsuruda, S. Tomishima, M. Tsukude, T. Yamagata, and K. Arimoto, "SOI-DRAM Circuit Technologies for Low Power High Speed Multigiga Scale Memories," *IEEE Journal of Solid-State Circuits*, vol. 31, pp. 586–591, April 1996.

[77] K. Shimomura, H. Shimano, N. Sakashita, F. Okuda, T. Oashi, Y. Yamaguchi, T. Eimori, M. Inuishi, K. Arimoto, S. Maegawa, Y. Inoue, S. Komori, and K. Kyuma, "A 1-V 46-ns 16-Mb SOI-DRAM with Body Control Technique," *IEEE Journal of Solid-State Circuits*, vol. 32, pp. 1712–1720, November 1997.

Embedded DRAM

[78] T. Sunaga, H. Miyatake, K. Kitamura, K. Kasuya, T. Saitoh, M. Tanaka, N. Tanigaki, Y. Mori, and N. Yamasaki, "DRAM Macros for ASIC Chips," *IEEE Journal of Solid-State Circuits*, vol. 30, pp. 1006–1014, September 1995.

Redundancy Techniques

[79] H. L. Kalter, C. H. Stapper, J. E. Barth, Jr., J. DiLorenzo, C. E. Drake, J. A. Fifield, G. A. Kelley, Jr., S. C. Lewis, W. B. van der Hoeven, and J. A. Yankosky, "A 50-ns 16-Mb DRAM with a 10-ns Data Rate and On-Chip ECC," *IEEE Journal of Solid-State Circuits*, vol. 25, pp. 1118–1128, October 1990.

[80] M. Horiguchi, J. Etoh, M. Aoki, K. Itoh, and T. Matsumoto, "A Flexible Redundancy Technique for High-Density DRAMs," *IEEE Journal of Solid-State Circuits*, vol. 26, pp. 12–17, January 1991.

[81] S. Kikuda, H. Miyamoto, S. Mori, M. Niiro, and M. Yamada, "Optimized Redundance Selection based on Failure-Related Yield Model for 64-Mb DRAM and Beyond," *IEEE Journal of Solid-State Circuits*, vol. 26, pp. 1550–1555, November 1991.

[82] T. Kirihata, Y. Watanabe, Hing Wong, J. K. DeBrosse, M. Yoshida, D. Kato, S. Fujii, M. R. Wordeman, P. Poechmueller, S. A. Parke, and Y. Asao, "Fault-

Tolerant Designs for 256 Mb DRAM," *IEEE Journal of Solid-State Circuits*, vol. 31, pp. 558–566, April 1996.

DRAM Testing

[83] T. Ohsawa, T. Furuyama, Y. Watanabe, H. Tanaka, N. Kushiyama, K. Tsuchida, Y. Nagahama, S. Yamano, T. Tanaka, S. Shinozaki, and K. Natori, "A 60-ns 4-Mbit CMOS DRAM with Built-in Selftest Function," *IEEE Journal of Solid-State Circuits*, vol. 22, pp. 663–668, October 1987.

[84] P. Mazumder, "Parallel Testing of Parametric Faults in a Three-Dimensional Dynamic Random-Access Memory," *IEEE Journal of Solid-State Circuits*, vol. 23, pp. 933–941, August 1988.

[85] K. Arimoto, Y. Matsuda, K. Furutani, M. Tsukude, T. Ooishi, K. Mashiko, and K. Fujishima, "A Speed-Enhanced DRAM Array Architecture with Embedded ECC," *IEEE Journal of Solid-State Circuits*, vol. 25, pp. 11–17, February 1990.

[86] T. Takeshima, M. Takada, H. Koike, H. Watanabe, S. Koshimaru, K. Mitake, W. Kikuchi, T. Tanigawa, T. Murotani, K. Noda, K. Tasaka, K. Yamanaka, and K. Koyama, "A 55-ns 16-Mb DRAM with Built-in Self-test Function Using Microprogram ROM," *IEEE Journal of Solid-State Circuits*, vol. 25, pp. 903–911, August 1990.

[87] T. Kirihata, Hing Wong, J. K. DeBrosse, Y. Watanabe, T. Hara, M. Yoshida, M. R. Wordeman, S. Fujii, Y. Asao, and B. Krsnik, "Flexible Test Mode Approach for 256-Mb DRAM," *IEEE Journal of Solid-State Circuits*, vol. 32, pp. 1525–1534, October 1997.

[88] S. Tanoi, Y. Tokunaga, T. Tanabe, K. Takahashi, A. Okada, M. Itoh, Y. Nagatomo, Y. Ohtsuki, and M. Uesugi, "On-Wafer BIST of a 200-Gb/s Failed-Bit Search for 1-Gb DRAM," *IEEE Journal of Solid-State Circuits*, vol. 32, pp. 1735–1742, November 1997.

Synchronous DRAM

[89] T. Sunaga, K. Hosokawa, Y. Nakamura, M. Ichinose, A. Moriwaki, S. Kakimi, and N. Kato, "A Full Bit Prefetch Architecture for Synchronous DRAM's," *IEEE Journal of Solid-State Circuits*, vol. 30, pp. 998–1005, September 1995.

[90] T. Kirihata, M. Gall, K. Hosokawa, J.-M. Dortu, Hing Wong, P. Pfefferi, B. L. Ji, O. Weinfurtner, J. K. DeBrosse, H. Terletzki, M. Selz, W. Ellis, M. R. Wordeman, and O. Kiehl, "A 220-mm/sup2/, Four-and Eight-Bank, 256-Mb SDRAM with Single-Sided Stitched WL Architecture," *IEEE Journal of Solid-State Circuits*, vol. 33, pp. 1711–1719, November 1998.

Low-Voltage DRAMs

[91] K. Lee, C. Kim, D. Yoo, J. Sim, S. Lee, B. Moon, K. Kim, N. Kim, S. Yoo, J. Yoo, and S. Cho, "Low Voltage High Speed Circuit Designs for Giga-bit DRAMs," 1996 Symposium on VLSI Circuits, p. 104, June 1996.

[92] M. Saito, J. Ogawa, K. Gotoh, S. Kawashima, and H. Tamura, "Technique for Controlling Effective V_{TH} in Multi-Gbit DRAM Sense Amplifier," 1996 Symposium on VLSI Circuits, p. 106, June 1996.

[93] K. Gotoh, J. Ogawa, M. Saito, H. Tamura, and M. Taguchi, "A 0.9 V Sense-Amplifier Driver for High-Speed Gb-Scale DRAMs," 1996 Symposium on VLSI Circuits, p. 108, June 1996.

[94] T. Hamamoto, Y. Morooka, T. Amano, and H. Ozaki, "An Efficient Charge Recycle and Transfer Pump Circuit for Low Operating Voltage DRAMs," 1996 Symposium on VLSI Circuits, p. 110, June 1996.

[95] T. Yamada, T. Suzuki, M. Agata, A. Fujiwara, and T. Fujita, "Capacitance Coupled Bus with Negative Delay Circuit for High Speed and Low Power (10GB/s < 500mW) Synchronous DRAMs," 1996 Symposium on VLSI Circuits, p. 112, June 1996.

High-Speed DRAMs

[96] S. Wakayama, K. Gotoh, M. Saito, H. Araki, T. S. Cheung, J. Ogawa, and H. Tamura, "10-ns Row Cycle DRAM Using Temporal Data Storage Buffer Architecture," 1998 Symposium on VLSI Circuits, p. 12, June 1998.

[97] Y. Kato, N. Nakaya, T. Maeda, M. Higashiho, T. Yokoyama, Y. Sugo, F. Baba, Y. Takemae, T. Miyabo, and S. Saito, "Non-Precharged Bit-Line Sensing Scheme for High-Speed Low-Power DRAMs," 1998 Symposium on VLSI Circuits, p. 16, June 1998.

[98] S. Utsugi, M. Hanyu, Y. Muramatsu, and T. Sugibayashi, "Non-Complimentary Rewriting and Serial-Data Coding Scheme for Shared-Sense-Amplifier Open-Bit-Line DRAMs," 1998 Symposium on VLSI Circuits, p. 18, June 1998.

[99] Y. Sato, T. Suzuki, T. Aikawa, S. Fujioka, W. Fujieda, H. Kobayashi, H. Ikeda, T. Nagasawa, A. Funyu, Y. Fujii, K. I. Kawasaki, M. Yamazaki, and M. Taguchi, "Fast Cycle RAM (FCRAM); a 20-ns Random Row Access, Pipe-Lined Operating DRAM," 1998 Symposium on VLSI Circuits, p. 22, June 1998.

High-Speed Memory Interface Control

[100] S.-J. Jang, S.-H. Han, C.-S. Kim, Y.-H. Jun, and H.-J. Yoo, "A Compact Ring Delay Line for High Speed Synchronous DRAM," 1998 Symposium on VLSI Circuits, p. 60, June 1998.

[101] H. Noda, M. Aoki, H. Tanaka, O. Nagashima, and H. Aoki, "An On-Chip Timing Adjuster with Sub-100-ps Resolution for a High-Speed DRAM Interface," 1998 Symposium on VLSI Circuits, p. 62, June 1998.

[102] T. Sato, Y. Nishio, T. Sugano, and Y. Nakagome, "5GByte/s Data Transfer Scheme with Bit-to-Bit Skew Control for Synchronous DRAM," 1998 Symposium on VLSI Circuits, p. 64, June 1998.

[103] T. Yoshimura, Y. Nakase, N. Watanabe, Y. Morooka, Y. Matsuda, M. Kumanoya, and H. Hamano, "A Delay-Locked Loop and 90-Degree Phase Shifter for 800Mbps Double Data Rate Memories," 1998 Symposium on VLSI Circuits, p. 66, June 1998.

High-Performance DRAM

[104] T. Kono, T. Hamamoto, K. Mitsui, and Y. Konishi, "A Precharged-Capacitor-Assisted Sensing (PCAS) Scheme with Novel Level Controlled for Low Power DRAMs," 1999 Symposium on VLSI Circuits, p. 123, June 1999.

[105] H. Hoenigschmid, A. Frey, J. DeBrosse, T. Kirihata, G. Mueller, G. Daniel, G. Frankowsky, K. Guay, D. Hanson, L. Hsu, B. Ji, D. Netis, S. Panaroni, C. Radens, A. Reith, D. Storaska, H. Terletzki, O. Weinfurtner, J. Alsmeier, W. Weber, and M. Wordeman, "A $7F^2$ Cell and Bitline Architecture Featuring Tilted Array Devices and Penalty-Free Vertical BL Twists for 4Gb DRAM's" 1999 Symposium on VLSI Circuits, p. 125, June 1999.

[106] S. Shiratake, K. Tsuchida, H. Toda, H. Kuyama, M. Wada, F. Kouno, T. Inaba, H. Akita, and K. Isobe, "A Pseudo Multi-Bank DRAM with Categorized Access Sequence," 1999 Symposium on VLSI Circuits, p. 127, June 1999.

[107] Y. Kanno, H. Mizuno, and T. Watanabe, "A DRAM System for Consistently Reducing CPU Wait Cycles," 1999 Symposium on VLSI Circuits, p. 131, June 1999.

[108] S. Perissakis, Y. Joo, J. Ahn, A. DeHon, and J. Wawrzynek, "Embedded DRAM for a Reconfigurable Array," 1999 Symposium on VLSI Circuits, p. 145, June 1999.

[109] T. Namekawa, S. Miyano, R. Fukuda, R. Haga, O. Wada, H. Banba, S. Takeda, K. Suda, K. Mimoto, S. Yamaguchi, T. Ohkubo, H. Takato, and K. Numata, "Dynamically Shift-Switched Dataline Redundancy Suitable for DRAM Macro with Wide Data Bus," 1999 Symposium on VLSI Circuits, p. 149, June 1999.

[110] C. Portmann, A. Chu, N. Hays, S. Sidiropoulos, D. Stark, P. Chau, K. Donnelly, and B. Garlepp, "A Multiple Vendor 2.5-V DLL for 1.6-GB/s RDRAMs," 1999 Symposium on VLSI Circuits, p. 153, June 1999.

Glossary

1T1C A DRAM memory cell consisting of a single MOSFET access transistor and a single storage capacitor.

Bitline Also called a digitline or columnline. A common conductor made from metal or polysilicon that connects multiple memory cells together through their access transistors. The bitline is ultimately used to connect memory cells to the sense amplifier block to permit Refresh, Read, and Write operations.

Bootstrapped Driver A driver circuit that employs capacitive coupling to boot, or raise up, a capacitive node to a voltage above V_{CC}.

Buried Capacitor Cell A DRAM memory cell in which the capacitor is constructed below the digitline.

Charge Pump *See* Voltage Pump.

CMOS, Complementary Metal-Oxide Semiconductor A silicon technology for fabricating integrated circuits. *Complementary* refers to the technology's use of both NMOS and PMOS transistors in its construction. The PMOS transistor is used primarily to pull signals toward the positive power supply V_{DD}. The NMOS transistor is used primarily to pull signals toward ground. The *metal-oxide semiconductor* describes the sandwich of metal oxide (actually polysilicon in modern devices) and silicon that makes up the NMOS and PMOS transistors.

COB, Capacitor over Bitline A DRAM memory cell in which the capacitor is constructed above the digitline (bitline).

Columnline *See* Bitline.

Column Redundancy The practice of adding spare digitlines to a memory array so that defective digitlines can be replaced with nondefective digitlines.

DCSA, Direct Current Sense Amplifier An amplifier connected to the memory array I/O lines that amplifies signals coming from the array.

Digitline *See* Bitline.

DLL, Delay-Locked Loop A circuit that generates and inserts an optimum delay to temporarily align two signals. In DRAM, a DLL synchronizes the input and output clock signals of the DRAM to the I/O data signals.

DRAM, Dynamic Random Access Memory A memory technology that stores information in the form of electric charge on capacitors. This technology is considered dynamic because the stored charge degrades over time due to leakage mechanisms. The leakage necessitates periodic Refresh of the memory cells to replace the lost charge.

Dummy Structure Additional circuitry or structures added to a design, most often to the memory array, that help maintain uniformity in live circuit structures. Nonuniformity occurs at the edges of repetitive structures due to photolithography and etch process limitations.

EDO DRAM, Extended Data Out DRAM A variation of fast page mode (FPM) DRAM. In EDO DRAM, the data outputs remain valid for a specified time after $\overline{CAS}$ goes HIGH during a Read operation. This feature permits higher system performance than would be obtained otherwise.

Efficiency A design metric, which is defined as the ratio of memory array die area and total die area (chip size). It is expressed as a percentage.

Equilibration Circuit A circuit that equalizes the voltages of a digitline pair by shorting the two digitlines together. Most often, the equilibration circuit includes a bias network, which helps to set and hold the equilibration level to a known voltage (generally $V_{CC}/2$) prior to Sensing.

Feature Size Generally refers to the minimum realizable process dimension. In the context of DRAM design, however, feature size equates to a dimension that is half of the digitline or wordline layout pitch.

Folded DRAM Array Architecture A DRAM architecture that uses non-crosspoint-style memory arrays in which a memory cell is placed only at alternating wordline and digitline intersections. Digitline pairs, for connection to the sense amplifiers, consist of two adjacent digitlines from a single memory array. For layout efficiency, each sense amplifier connects to two adjacent memory arrays through isolation transistor pairs.

FPM, Fast Page Mode A second-generation memory technology permitting consecutive Reads from an open page of memory, in which the column address could be changed while $\overline{CAS}$ was still low.

Glossary

Helper Flip-Flop, HFF A positive feedback (regenerative) circuit for amplifying the signals on the I/O lines.

I/O Devices MOSFET transistors that connect the array digitlines to the I/O lines (through the sense amplifiers). Read and Write operations from/to the memory arrays always occur through I/O devices.

Isolation Devices MOSFET transistors that isolate array digitlines from the sense amplifiers.

Mbit, Memory Bit A memory cell capable of storing one bit of data. In modern DRAMs, the mbit consists of a single MOSFET access transistor and a single storage capacitor. The gate of the MOSFET connects to the wordline or rowline, while the source and drain of the MOSFET connect to the storage capacitor and the digitline, respectively.

Memory Array An array of memory or mbit cells.

Multiplexed Addressing The practice of using the same chip address pins for both the row and column addresses. The addresses are clocked into the device at different times.

Open DRAM Array Architecture A DRAM architecture that uses cross-point-style memory arrays in which a memory cell is placed at every wordline and digitline intersection. Digitline pairs, for connection to the sense amplifiers, consist of a single digitline from two adjacent memory arrays.

Pitch The distance between like points in a periodic array. For example, digitline pitch in a DRAM array is the distance between the centers or edges of two adjacent digitlines.

RAM, Random Access Memory Computer memory that allows access to any memory location without restrictions.

Refresh The process of restoring the electric charge in DRAM memory cell capacitors to full levels through Sensing. Note that Refresh occurs every time a wordline is activated and the sense amplifiers are fired.

Rowline *See* Wordline.

SDRAM, Synchronous DRAM A DRAM technology in which addressing, command, control, and data operations are accomplished in synchronism with a master clock signal.

Sense Amplifier A type of regenerative amplifier that senses the contents of memory cells and restores them to full levels.

Trench Capacitor A DRAM storage capacitor fabricated in a deep hole (trench) in the semiconductor substrate.

Voltage Pump Also called a charge pump. A circuit for generating voltages that lie outside of the power supply range.

Wordline Also called a rowline. A wordline is a polysilicon conductor for forming memory cell access transistor gates and connecting multiple memory cells into a physical row. Driving a wordline HIGH activates, or turns ON, the access transistors in a memory array row.

Index

Numerics
1T1C memory cell, 22, 35
3-transistor DRAM cell, 7

A
Access time (t_{AC}), 4, 21, 142
ACT. See Active pull-up
Active pull-up *(ACT)*, 28–29, 52
Address
    decode trees, 62
    multiplexing, 4
    path elements, 132
    predecode, 64, 114, 140
    transition detection, 139
Address to Write delay time (t_{AW}), 5
arbiter, 149
Array, 24
    architecture, 69
    architecture, comparison, 98
    buffers, 137
    efficiency, 74, 76–77, 81, 84, 98–99

B
Bandgap reference, 161
Bias circuits, 46
Bilevel
    digitline, 89
    digitline array architecture, 93
Bitline, 10
    capacitance number, limitations, 12
    PRECHARGE voltage, 14
Bitline over capacitor (BOC), 35
BOC. *See* Bitline over capacitor
BOOST signal, 165
Bootstrap wordline driver, 58
Breakdown voltages, 155
Buried capacitor, 35, 41
Buried digitline, 42
Burn-in, 157

C
Capacitor
    buried, 43
    trench, 45
Capacitor over bitline (COB), 42
Capacity breakdown, 42

$\overline{CAS}$. *See* Column address strobe
C_{digit}, digitline capacitance, 27
Charge pump, 166
C_{IN}, input capacitance, 2
C_{mbit}, mbit capacitance, 27
CMOS operational amplifiers, 158
COB. *See* Capacitor over bitline
Column
    address path, 138
    decoder, 105
    redundancy, 105, 108, 113
Column address strobe $(\overline{CAS})$, 8, 18
Column hold time (t_{CAH}), 8
Column select *(CSEL)*, 30, 49
Column setup time (t_{ASC}), 8
Columnline, 10, 14
CSEL. *See* Column select

D

Data input buffer, 117
Data path elements, 117
Data Read
    muxes, 129
    path, 124
DC sense amplifier *(DCSA)*, 124–125
DCSA. *See* DC sense amplifier
DDR. *See* Double data rate
Delay-locked loop *(DLL)*, 142
Design examples, 87
Digitline, 10, 14, 24, 37
    capacitance, 27, 40
    construction, 41
    floating, 54
    open digitline array layout, 40
    pair, 26, 28
    twist, 89

    voltages, 51
DLL. *See* Delay-locked loop
Donut gate, 60
Double data rate (DDR), 17, 142
DQ, 14
    mask (DQM), 19
    strobe (DQS), 17
DQM. *See DQ* mask
DQS. *See DQ* strobe
*DVC*2, 155
Dynamic random access memory (DRAM), 1, 13
    1T1C memory cell, 10
    3-transistor cell, 7
    address multiplexing, 4
    array, 35
    evolution, 1
    first generation, 1
    internal V_{CC}, 156
    layout, 24
    mbit, 22
    modes of operation, 13
    opening a row, 11, 13–14, 18, 31
    organization, 12
    power consumption, 49
    power dissipation, 41, 96
    Read, 26
    reading data out, 4
    Refresh, 5, 13
    second generation, 7
    synchronization, 142
    synchronous, 16
    test modes, 131
    testing, 55
    types and operation, 1
    Write, 30

Index

E
EDO. *See* Extended data out
Embedded DRAM, 32
EQ. See Equilibrate
Equilibrate *(EQ)*, 47
Equilibration, 26
    and bias circuits, 46
Extended data out (EDO), 8, 14, 105

F
Fast page mode (FPM), 8, 14, 105
Folded array, 38
    architecture, 79
Folded digitline architectures, 69
FPM. *See* Fast page mode
Fully differential amplifier, 121

G
Global circuitry, 117

H
Helper flip-flop (HFF), 32, 124, 127
HFF. *See* Helper flip-flop
High-input trip point (V_{IH}), 118

I
I/O, 17
I/O transistors, 30, 49
Input buffer, 1, 119
Input capacitance (C_{IN}), 2
Isolation, 48
    array, 94
    transistors, 46

J
Jitter, 147

L
Leakage, 36
Low-input trip point (V_{IL}), 118

M
Mbit, 22
    $6F^2$, 38
    $8F^2$, 37
    bitline capacitor, 35
    buried capacitor, 41
    capacitance (C_{mbit}), 27
    layout, 24
Mbit pair layout, 36
Memory
    element, 2
Memory array, 2, 10–11
    layout, 2
    size, 12
Multiplexed addressing, 8

N
N sense-amp latch *(NLAT*)*, 28, 52
Nibble mode, 8, 14–15
NLAT*. See N sense-amp latch

O
ONO dielectric. *See* Oxide-nitride-oxide dielectric
Open architectures, 69
Open array, 38
Open digitline array, 69
Opening a row, 11–14, 18, 31
Oscillator circuits, 168
Output buffer circuit, 129
Oxide-nitride-oxide (ONO) dielectric, 41

P

Page
 mode, 8, 14
 Reads, 11
 size, 13
PD. *See* Phase detector
Peripheral circuitry, 105
Phase detector *(PD)*, 144
Phase drivers, 137
Pitch, 46
 cells, 46
pnp transistors, 161
Polycide, 11
Power
 consumption, 49
 dissipation, 41, 96
POWERDOWN, 156
POWERUP, 156
PRECHARGE, 26, 47
Predecode logic, 135
Predecoding, 64
Pumps and generators, 166

R

$R/\overline{W}$, 5
$\overline{RAS}$. *See* Row address strobe
Rate of activation, 52
RDLL. *See* Register-controlled delay-locked loop
Read cycle time (t_{RC}), 4
Read-Modify-Write, 14, 56
Redundancy, 108
Refresh, 5, 11, 13, 28, 135
Register-controlled delay-locked loop (RDLL), 142

Row address strobe $(\overline{RAS})$, 8, 18, 47
Row decode, 57
Row hold time (t_{RAH}), 8
Row redundancy, 108, 110
Row setup time (t_{ASR}), 8
Rowline, 10, 24

S

Sense amplifier, 28, 117
 configurations, 52
 Nsense, Psense, 50
 operation, 55
Sensing, 32
Silicide, 11
SMD. *See* Synchronous mirror delay
SSTL. *See* Stub series terminated logic
Static column mode, 8, 14–15
Stub series terminated logic (SSTL), 119
Subthreshold leakage, 51
Synchronization in DRAMs, 142
Synchronous DRAM (SDRAM), 16, 105
 DDR, double data rate, 17, 142
Synchronous mirror delay (SMD), 142

T

t_{AC}. *See* Access time
t_{ASC}. *See* Column setup time
t_{ASR}. *See* Row setup time
t_{AW}. *See* Address to Write delay time
t_{CAH}. *See* Column hold time
Test modes, 131
t_{RAH}. *See* Row hold time
t_{RC}. *See* Read cycle time

Index **197**

Trench capacitor, 45
TTL, 118
    logic, 6
t_{WC}. *See* Write cycle time
Twisting, 37
t_{WP}. *See* Write pulse width

V

V_{BB} pump, 166
$V_{CC}/2$, 155
V_{CCP}, 10, 26, 29
V_{IH}. *See* High-input trip point
V_{IL}. *See* Low-input trip point
Voltage
    converters, 155
    pump, 166
    references, 156
    regulator characteristics, 159
    regulators, 155

W

Wordline, 10, 24, 35
    CMOS driver, 61
    delay, 11, 26
    NOR driver, 60
    number, limitations, 11
    pitch, 36
Write cycle time (t_{WC}), 5
Write driver, 30
Write driver circuit, 122
Write operation waveforms, 31, 33
Write pulse width (t_{WP}), 5

Y

Yield, 1

About the Authors

Brent Keeth was born in Ogden, Utah, on March 30, 1960. He received the B.S. and M.S. degrees in electrical engineering from the University of Idaho, Moscow, in 1982 and 1996, respectively.

Mr. Keeth joined Texas Instruments in 1982, spending the next two years designing hybrid integrated circuits for avionics control systems and a variety of military radar subsystems. From 1984 to 1987, he worked for General Instruments Corporation designing baseband scrambling and descrambling equipment for the CATV industry.

Thereafter, he spent 1987 through 1992 with the Grass Valley Group (a subsidiary of Tektronix) designing professional broadcast, production, and post-production video equipment. Joining Micron Technology in 1992, he has engaged in the research and development of various CMOS DRAMs including 4Mbit, 16Mbit, 64Mbit, 128Mbit, and 256Mbit devices. As a Principal Fellow at Micron, his present research interests include high-speed bus protocols and open standard memory design.

In 1995 and 1996, Brent served on the Technical Program Committee for the Symposium on VLSI Circuits. In addition, he served on the Memory Subcommittee of the U.S. Program Committee for the 1996 and 1999 IEEE International Solid-State Circuits Conferences. Mr. Keeth holds over 60 U.S. and foreign patents.

R. Jacob Baker (S'83, M'88, SM'97) was born in Ogden, Utah, on October 5, 1964. He received the B.S. and M.S. degrees in electrical engineering from the University of Nevada, Las Vegas, and the Ph.D. degree in electrical engineering from the University of Nevada, Reno.

From 1981 to 1987, Dr. Baker served in the United States Marine Corps Reserves. From 1985 to 1993, he worked for E.G.&G. Energy Measurements and the Lawrence Livermore National Laboratory designing nuclear diagnostic instrumentation for underground nuclear weapons tests at the Nevada test site. During this time, he designed over 30 electronic and electro-optic instruments, including high-speed (750 Mb/s) fiber-optic receiver/transmitters, PLLs, frame- and bit-syncs, data converters, streak-camera sweep circuits, micro-channel plate gating circuits, and analog oscilloscope electronics. From 1993 to 2000, he was a faculty member in the Department of Electrical Engineering at the University of Idaho. In 2000, he joined a new electrical engineering program at Boise State University as an associate professor. Also, since 1993, he has consulted for various companies, including the Lawrence Berkeley Laboratory, Micron Technology, Micron Display, Amkor Wafer Fabrication Services, Tower Semiconductor, Rendition, and the Tower ASIC Design Center.

Holding 12 patents in integrated circuit design, Dr. Baker is a member of Eta Kappa Nu and is a coauthor (with H. Li and D. Boyce) of a popular textbook covering CMOS analog and digital circuit design entitled, *CMOS: Circuit Design, Layout, and Simulation* (IEEE Press, 1998). His research interests focus mainly on CMOS mixed-signal integrated circuit design.